Metabolomics by *In Vivo* NMR

Metabolomics by *In Vivo* NMR

Editors

R G Shulman and D L Rothman

Yale University School of Medicine
New Haven, Connecticut, USA

John Wiley & Sons, Ltd

Telephone (+44) 1243 779777

Email (for orders and customer service enquiries): cs-books@wiley.co.uk
Visit our Home Page on www.wileyeurope.com or www.wiley.com

Other Wiley Editorial Offices

John Wiley & Sons Inc., 111 River Street, Hoboken, NJ 07030, USA

Jossey-Bass, 989 Market Street, San Francisco, CA 94103-1741, USA

Wiley-VCH Verlag GmbH, Boschstr. 12, D-69469 Weinheim, Germany

John Wiley & Sons Australia Ltd, 33 Park Road, Milton, Queensland 4064, Australia

John Wiley & Sons (Asia) Pte Ltd, 2 Clementi Loop #02-01, Jin Xing Distripark, Singapore 129809

John Wiley & Sons Canada Ltd, 22 Worcester Road, Etobicoke, Ontario, Canada M9W 1L1

Wiley also publishes its books in a variety of electronic formats. Some content that appears in print may not be available in electronic books.

Library of Congress Cataloging-in-Publication Data

Metabolomics by in vivo NMR / editors, R.G. Shulman and D.L. Rothman.
p. ; cm.
Includes bibliographical references and index.
ISBN 0-470-84719-0 (cloth : alk. paper)
1. Metabolism. 2. Metabolism – Research – Methodology. 3. Nuclear magnetic resonance spectroscopy.
[DNLM: 1. Metabolism – physiology. 2. Cell Physiology. 3. Magnetic Resonance Spectroscopy – methods. 4. Muscles – metabolism. QU 120 M587 2005] I. Title: Metabolomics by in vivo nuclear magnetic resonance. II. Shulman, R. G. (Robert Gerson) III. Rothman, D. L. (Douglas L.)
QP171.M383 2005
572′.41′072 – dc22

2004013115

British Library Cataloguing in Publication Data

A catalogue record for this book is available from the British Library

ISBN 0-470-84719-0

Typeset in 10/12pt Times by Laserwords Private Limited, Chennai, India
Printed and bound in Great Britain by Antony Rowe Ltd, Chippenham, Wiltshire
This book is printed on acid-free paper responsibly manufactured from sustainable forestry in which at least two trees are planted for each one used for paper production.

This volume is dedicated to Lionel Trilling who taught us the excitement of thought.

‘The fascination of a growing science lies in the work of the pioneers at the very borderland of the unknown, but to reach this frontier one must pass over well traveled roads; of these one of the safest and surest is the broad highway of thermodynamics’, Lewis, G.N. and Randall, M. (1923) *Thermodynamics and the Free Energy of Chemical Substances*, McGraw-Hill, New York.

Contents

Contributors

Robin de Graaf
Department of Diagnostic Radiology
Yale University School of Medicine
MR Center, PO Box 208043
New Haven, CT 06520-8043, USA
robin.degraaf@yale.edu

Jan den Hollander
University of Alabama at Birmingham
Center for NMR Research and Development
Birmingham, AL 35294, USA

David A. Fell
School of Biological and Molecular Sciences
Oxford Brookes University
Gipsy Lane
Headington, Oxford OX3 0BP, UK
dfell@bms.brookes.ac.uk

Thomas Jue
Medicine: Biological Chemistry
University of California, Davis
Davis, CA 95616, USA
tjue@ucdavis.edu

Maren R. Laughlin
National Institute of Diabetes and Digestive
and Kidney Diseases
6707 Democracy Blvd, Room 6101, MSC 5460
Bethesda, MD 20892-5460, USA
laughlinm@extra.niddk.nih.gov

Craig R. Malloy
UT Southwestern Medical Center
Department of Radiology
Dallas, TX 75235-9085, USA
Craig.Malloy@UTSouthwestern.edu

Graeme Mason
Department of Psychiatry
Yale University School of Medicine
MR Center, PO Box 208043
New Haven, CT 06520-8043, USA
graeme.mason@yale.edu

Thomas B. Price
Yale University School of Medicine
Department of Diagnostic Radiology
PO Box 208042
New Haven, CT 06520-8042, USA
price@boreas.med.yale.edu

Sir George Radda
Medical Research Council
20 Park Crescent
London W1N 4AL, UK

Douglas L. Rothman
Department of Diagnostic Radiology
Yale University School of Medicine
MR Center, PO Box 208043
New Haven, CT 06520-8043, USA
douglas.rothman@yale.edu

James R.A. Schafer
Department of Neuroscience
Yale University School of Medicine
MR Center, PO Box 208043
New Haven, CT 06520-8043, USA
james.schafer@yale.edu

A. Dean Sherry
University of Texas at Dallas
Richardson, TX 75083, USA
dean.sherry@Southwestern.edu

Gerald I. Shulman
Internal Medicine Section of Endocrinology
Yale University School of Medicine
PO Box 208056
New Haven, CT 06520-8056, USA
gerald.shulman@yale.edu

Robert G. Shulman
Yale University School of Medicine
Department of Diagnostic Radiology
PO Box 208024
New Haven, CT 06520-8024, USA
robert.shulman@yale.edu

Foreword

Bob Shulman and I were fortunate to be involved in the early developments of *in vivo* NMR. The field was beginning to open up, full of exciting promises, and competition between our two laboratories was healthy and desirable. At the same time during our friendly discussions we began to realize that our philosophies about the nature of this new approach were very similar and our experimental studies were complementary. For this reason, in 1984 we decided to write a joint contribution to a book, *Biomedical Magnetic Resonance*, edited by Tom James and Alex Margulis. Our article, 'Nuclear magnetic resonance of *in vivo* metabolism: from normal to pathophysiology', was published just 10 years after the first 31-phosphorus NMR study of intact isolated rat muscle and several papers from the Shulman laboratory on 13-carbon NMR studies on intact yeast cells. In our article we outlined 'how parallel and complementary developments in NMR spectroscopy in our two laboratories have helped to derive rich biochemical information from spectroscopic studies and to show how this information is beginning to have medical applications.' We noted that 'chemical energy is provided to the living cell by two primary biochemical pathways: glycolysis and oxidative phosphorylation' and that 'textbook description of these two pathways is likely to be largely correct, in contrast our ideas about molecular control of these events and the way they are co-ordinated with other functions such as energy demand, are largely speculative.' We also remarked that 'this is not surprising as understanding control requires that we should be able to study reaction fluxes, molecular interactions and metabolite distributions within the intact organism.'

Since 1984 the field has expanded enormously and our discussions became more wide-ranging and our friendship grew stronger. During a long walk in the woods surrounding the Max Grundig Clinic in Bühlerhöhe near Baden-Baden in 1994 Bob outlined his ideas for this book. I am delighted that the book on *Metabolomics by In vivo NMR* is now a reality. However, I am also pleased that it was not written earlier. The spectacular advances in genome sequencing and genomics in general in the last five or so years made it even more imperative that molecular networks, interactions and intracellular reaction fluxes be studied in the imaginative way that Bob Shulman and his colleagues developed the use of cellular NMR spectroscopy. I am therefore pleased about the timing of the book as it will have a more significant impact on future thinking about *in vivo* NMR as one of the important tools of the 'post-genomic' era. We not only need to understand cellular functions at the molecular level but also need to be able to bring the knowledge gained to solve problems in human disease. *In vivo* NMR spectroscopy and other forms of molecular imaging will bring clinical practice and biochemical understanding closer together. An example of this is the work reported by Bob in Bühlerhöhe under the title 'Nuclear Magnetic resonance studies of muscle and applications to exercise and diabetes' [*Diabetes* (1966) **45** (Suppl. 1): S93–S98]. This theme is developed further in this book by Jerry Shulman, who shows how fundamental understanding of control of glycogenolysis and glucose metabolism in skeletal muscle could contribute to the clinical management of diabetes.

The DNA in the genome is the depository of biological information. Another level of cellular organization is responsible for the processing of this information and the 'metabolome', as emphasized in this book, provides the basis for the execution of the various cellular programmes. However, it is becoming increasingly evident that there is a complex network of interactions between these different organizational levels. The metabolic influence on transcription factor-controlled information processing is just one further piece of evidence for the need to look at cellular functions in a highly integrated way. A striking example of such interactions comes from the recent introduction of a new class of insulin-sensitizing drugs, the thiazolidines, that act as antagonists of the peroxisome proliferator activated receptor (PPARy), which is involved in the regulation of several transcription factors, including one for the expression of one of the mitochondrial uncoupling proteins. Serum free fatty acids (FFA) interact with this receptor. Thus understanding the relationship between obesity, diabetes, regulation of gene expression and energy provision and utilization is within our grasp. Biology has been transformed from an observational, descriptive science to a quantitative science. *In vivo* NMR as described in this book is on the path to this New Biology.

George K. Radda
Medical Research Council,
London, UK January 2003

1

Introduction

Robert G. Shulman

Yale University School of Medicine, Department of Diagnostic Radiology, PO Box 208043, New Haven, CT 06520-8024, USA

The wonderful advances of biochemistry, molecular biology and structural biology in the last half-century have created a foundation which *in vivo* NMR has utilized in the attack upon more complex organismic levels. Metabolic pathways, whose fluxes and intermediates are controlled by enzymes and genes in response to an organism's needs, are considered by some to be a worn-out topic of investigation – one that is largely 'complete'. The basic concepts are well established, the argument goes, and only minor details remain to be worked out. We feel that this view misses the most important point of reductionist biology – namely, that which has been established at one level can illuminate the more complex. For the past three decades scientists using *in vivo* nuclear magnetic resonance spectroscopy (MRS) to study metabolism have been developing an interdisciplinary vision for the future of biology at the molecular level. While their experimental findings and the immediate applications to particular metabolic pathways have been published extensively, the overall goals and philosophy of these individual studies have lurked, undeclared, below the scientific expositions. Only when these reports are considered together can one see that they build stepwise upon one another and ultimately provide a coherent direction for the future of biochemistry. We have had the pleasure of taking part in many of these studies and intend this book to serve as a review that outlines the theoretical framework that has been present in our minds but which has remained only implicit in our publications. Reviews are never complete and the progress they describe invariably changes, but we nevertheless are encouraged to assemble our report at this time. Our excitement stems from the depth and breadth of the results we wish to present. In the past 20 years many have asked, 'When are the biological results going to be worthy of the novel experimental methods developed for *in vivo* MRS?' In reviewing the literature for this book we are convinced that, to a communicable extent, the goal has been realized so that the use of *in vivo* MRS for the clarification of molecular events is well established. In addition to our enthusiasm, we are also motivated by a sense of obligation that arises from our conviction that biochemistry sits at a historic crossroads. Down one, now heavily traveled, path we see overwhelmed researchers chained to vast computers into which they feed ever-larger collections of descriptive data in the paradoxical hope that the only cure for the current information deluge is an

Metabolomics by In Vivo NMR. Edited by R. G. Shulman and D. L. Rothman
 ISBN: 0-470-84719-0

exponential increase in the number of data points. Down the other, less crowded, are researchers directly examining dynamic metabolic systems that use the genes, proteins and metabolites to realize biological functions. This book, in examining results from *in vivo* MRS studies of humans and animals, outlines ways in which this second path has been taken, and suggests how they might be applied more broadly to both experiments and ideology.

Practically, we propose that biochemistry can now reliably move upwards in complexity towards systemic physiology and beyond by making *in vivo* MRS studies of metabolic pathways. Biochemistry, like most biological fields, is intermediate between the reductionist explanations offered by chemistry and physics and the organismic functions studied by medicine, behavior or evolution. Unfortunately, it is much easier to simply collect more information about one particular level of complexity than to make the transition to the next. To evaluate such a relationship requires dynamic interplay between observations, new methodologies and entirely new concepts.

Biochemists are increasingly calling for more organismic insights – for 'functional genomics' and 'proteomics'. These visions, and the parallel medical hopes for genetic determinism, are in fact running headlong into the complexities of metabolism. Just as 'gene products' is a new word for enzymes, 'functional genomics', describing the functions served by gene products, is a new word for metabolism. In our view, *in vivo* MRS offers more informative ways of studying the metabolic pathways, and thus of moving from the relatively well-understood enzymes and metabolites to the organism-level phenomena of physiology and behavior.

In vivo MRS offers the opportunity to directly attach questions in functional genomics/metabolism, instead of pursuing these questions in a bottom-up (genetic) or top-down (cell biology) approach. 'Metabolomics' is a term with varied meanings but which most generally describes a large-scale, organized scientific attack upon the complexities of metabolism. It recognizes the importance of metabolism, and seeks to consolidate existing and newly generated information about metabolic pathways and constituents in the same way that proteomics endeavors to catalog biomolecular structures and protein expression profiles, and genomics addresses the vast arrays of DNA sequence information. In one generally accepted formulation, metabolomics plans to identify all the molecular components in the cell, including metabolites, macromolecules, cellular structures and solution conditions. It plans to use information available from proteomics and genomics to develop insights into organismic complexity. This book shares that goal but proposes that it can be pursued with MRS studies of metabolic fluxes *in vivo*, and with narrower but more realizable goals in mind (e.g. to understand organismic function at the level of physiology). By studying selected pathways with an eye for general principles, one can relate specific biochemical knowledge to insights about physiological processes. One result emerging in the following chapters is that, the more we understand the control of biochemical metabolism, the more clearly we understand how biochemical metabolism subserves physiological needs, how dependent biochemical control is upon physiology.

The unique perspective of this book derives from the strengths of ^{13}C NMR, a technique with the ability to measure metabolic fluxes in living humans and animals. *In vivo* methods for ^{13}C NMR detection are extensions of earlier developments in organic chemistry in which the resonance (or chemical shift) of the ^{13}C peak identifies a particular carbon of a particular molecule. For these assignments, sharp well-resolved resonances are required as observed for small molecules (MW $< 10^3$) and occasionally in larger molecules such as glycogen, where there are rapid internal motions. The natural abundance of the ^{13}C isotope is 1.1 % and it is NMR-visible, while the other $\sim$99 % of the carbon nuclei are not NMR-visible. The low natural abundance allows two kinds of measurements: the first in which the ^{13}C NMR intensity allows the concentration of the compound to be calculated; and the second, which follows from increasing the ^{13}C fractional enrichment by introducing substrates, such as 1-^{13}C glucose, in which the ^{13}C enrichment has been increased towards 100 %. In this case, as the substrate reacts, the label moves down the biochemical pathway to subsequent compounds. From this measurement the flux through the sequence of reactions (i.e.

the pathway) can be calculated from the ^{13}C NMR intensities measured either as a function of time or measured when a steady state has been reached. These measurements provide the quantitative assessments of metabolite concentrations and rates of reaction, *in vivo*, which provide the unique, strong data sets that have allowed metabolism, studied this way, to provide a platform for novel insights into biological function.

In addition to the experimental opportunities offered by *in vivo* MRS, the study of metabolic pathways has been enriched by the development of metabolic control analysis (MCA). This analytic system provides a model in which *in vivo* MRS measurements of pathway flux and *in vitro* measurements of enzyme kinetics can be combined to make quantitative evaluations of the control of flux and of the concentrations of intermediates. The formal quantitative and theoretical aspects of MCA have been simplified in the past two decades so as to be readily applicable to *in vivo* MRS experiments. The essential simplified features of MCA needed to interpret the experiments discussed are presented in this book.

Three chapters introduce the book – this first chapter presents a unifying overview of the following sections. It delineates the roles proposed for *in vivo* MRS studies in physiology and medicine as interpreted and guided by MCA, and emphasizes the symbiotic strengths obtained by combining these two approaches. The second chapter summarizes the *in vivo* NMR techniques used in the subsequent chapters to derive the *in vivo* concentrations of metabolites and pathways fluxes. Although necessarily technical in nature, the explanations were written to be easily accessible to an undergraduate chemistry student. Chapter 3 outlines MCA, defining terms and presenting relations to be applied in several subsequent studies. MCA was invented more than 30 years ago, mainly by Kacser in Edinburgh, to explain the role of mutation in microorganisms. It defines the control of flux and the regulation of metabolite concentrations in terms of the measurable properties of the constituent metabolites and enzymes in a pathway. It represents a particularly valuable achievement of metabolomics, by deriving a functional role of these constituents *in vivo*. In establishing criteria for the control of flux by each enzyme in a pathway, it gives quantitative meaning to the concept of flux control, a concept whose differing definitions have led to confusion. The definitions and the parameters of MCA have been expressed as partial derivatives. This allows for valuable quantitation but, in light of the growing view that mathematics conceals knowledge, has undoubtedly has been responsible for MCA's neglect by biochemists. *In vivo* MRS experiments, which determine quantitative values of pathway fluxes and metabolite concentrations, are ideally suited to providing the sort of data MCA can then use to generate a coherent picture of metabolic systems. The simple formulations of MCA needed to take advantage of the MRS results are described in Chapter 3.

Chapters 4– 7 describe a metabolic system extensively studied by ^{13}C NMR and which focuses on glycogen synthesis, a valuable perspective for studying metabolic control in activity and disease. Chapter 4 reviews and integrates studies of glucose metabolism with a focus on the role of glycogen. These studies build upon the 100 % visibility of the ^{13}C resonances of this very large molecule – which allows concentrations and rates of synthesis to be obtained *in vivo* from the NMR spectra. The rate of glycogenesis in muscle as measured in non-insulin-dependent diabetic mellitus (NIDDM) subjects and their matched normal controls is described. The experiments provide understanding of the metabolic basis of glycogenesis and of the defects in NIDDM. They show that glucose after a meal is predominantly stored as muscle glycogen, that the rate of this pathway is reduced in NIDDM (thus explaining hyperglycemia) and that this reduced rate in NIDDM is caused by reduced recruitment of glucose transporters to the plasma membrane in response to insulin. This led to questions about the role of the enzyme glycogen synthase (GSase), the canonical example in biochemistry textbooks of an allosteric enzyme whose activity is under the control of a phosphorylation cascade. GSase, which is often considered to control the flux of glycogen synthesis as a result of its extensive modulation, is shown in Chapter 5 to not control this flux. Instead, Chapter 5 shows that GSase serves to maintain homeostasis, keeping the metabolic intermediates concentrations close to constant despite flux increases. Homeostasis is a physiological parameter, needed by other functioning elements in the body, so that this pathway property relates the flux through a pathway to systemic functions.

Furthermore the flux of glycogen synthesis itself serves homeostasis by maintaining blood glucose levels constant and storing the excess.

These results on the particular pathway of glycogenesis have implications that extend far beyond the specific pathway of muscle glycogen synthesis. The findings are relevant for other functional aspects of enzymes studied by modern biochemistry. Signaling pathways often express kinases, enzymes generally assumed to control fluxes by phosphorylating 'rate-limiting' enzymes in pathways. However, the combined function of allostery and phosphorylation in the archetypical phosphorylated enzyme, GSase, is shown not to be flux control of glycogen synthesis. Rather, it is to maintain metabolite concentration constant during changes in the flux of glycogen synthesis. This novel function for phosphorylation is an alternative to its generally accepted role of flux control, and is a broadly applicable finding of *in vivo* MRS as interpreted by MCA.

Chapter 6 describes muscle glycogen measurements during exercise. Topics include the depletion and restoration of glycogen stores during intense anaerobic exercise and moderate long-term aerobic exercise. Measurements of the change in total glycogen concentration in response to difference states of exercise were made by quantifying the natural abundance of ^{13}C NMR peaks. A particularly valuable measurement was made by infusing labeled 1-^{13}C glucose at a constant glycogen concentration in order to observe how different parameters affect glycogen turnover. The continual synthesis and consumption of glycogen during prolonged exercise has reinforced the notion that glycogen serves to supply the rapid pulse of energy needed during muscle twitches.

Chapter 7 reviews data on the synthesis and degradation of glycogen in the heart. It compares the results of ^{13}C MRS studies of myocardial glycogen metabolism with the skeletal muscle results, indicating similarities and differences. Heart data show that glucose uptake seems to be controlled by the glucose transporter/hexokinase entry step, as in muscle. Differences occur when the metabolic pathway bifurcates at G6P where the glycolytic flux is more active in the heart due to its ceaseless activity. The competition between these two paths for glucose flux and its influence by hormones and non-glucose substrates provide a recurring comparison between cardiac and skeletal muscle.

The next two chapters follow the existence and consequences of the rapid energy metabolism needed to support muscle contractions. Chapter 8 describes gated ^{31}P NMR measurements of phosphocreatine (PCr) in muscle. These data were acquired with a millisecond time resolution, so they allow for evaluation of the dynamic nature of muscle energetics. Specifically, by measuring the fall and rise of PCr during muscle contraction they provide a real-time basis for the energetics of muscle contractions. The usual values of energy consumption are obtained by comparing ^{31}P NMR measurements of PCr concentrations before and after a period of exercise lasting seconds or minutes. The gated NMR studies described here show how that approach severely underestimates the energy consumption because the PCr is substantially depleted and regenerated in synchrony with the contractions. The possibility that the rapid energy production must be provided by the anaerobic pathway leads to Chapter 9, which examines the role of lactate, the product of anaerobic glycolysis. The results disagree with the traditional views in which lactate is produced because of insufficient oxygen. Recent data have shown lactate appearances in well-oxygenated muscle, while the many recent functional magnetic resonance imaging results in the brain also show that lactate is produced during stimulation in well-oxygenated tissue. In Chapter 9 it is proposed that lactate is produced by the rapid anaerobic processes that respond to contraction. In this model lactate serves as a temporal buffer relating the millisecond need for energy to the long-term measurements of PCr glucose, glycogen and oxygen consumption.

In vivo ^{13}C and ^{31}P measurements were made on cellular suspensions several years before it was technically possible to study animal models and humans. Very detailed studies of energetic pathways in yeast and yeast spores made 20 years ago are juxtaposed with recent related studies in Chapters 10 and 11, and seen to provide novel answers to contemporary questions about metabolic adaptation to environment.

Differences and similarities of control mechanisms in yeast, described in Chapter 10, with activities of the same paths of glucose metabolism in human muscle, show the versatility of these shared energetic pathways. Central to the comparison is that, similar to mammalian pathways, which are shown in earlier chapters to be subservient to the organism's physiological needs, yeast can be subjected to rapidly changing environments (e.g. the sudden infusion of glucose, or the loss of nutrients) and has thus evolved mechanisms of survival despite such perturbations. In conformity with mammalian systems, yeast and yeast spores depend upon efficient energy usage during steady-state periods and are particularly well adapted to the cellular needs during transient periods when environments change rapidly.

Chapter 12 illustrates the great usefulness of isotopomer analysis. This method takes advantage of the fact that nearby ^{13}C sites in metabolites give spin-coupled NMR spectra. Labels at each site reflect the accumulated flux through different pathways so that measurements of the coupling can determine fluxes. This method does not require *in vivo* time course measurements and can be made at steady state either *in vivo* or, often, on extracts. Isotopomer experiments are made on body fluids or extracts with high-resolution NMR spectrometers. Their vertical magnets and rapid spectroscopic measurements are readily adapted to high throughput. Large data sets can be obtained and the specificity of metabolic flux determinations extended to large populations.

The last chapter summarizes results from the previous sections and discusses their relationship to general questions relevant to biological studies and to the goals implicit in metabolomics. The rapidly changing understanding of biochemistry and physiology based upon the new experimental approaches and guided by the formal structures of MCA, moves us confidently from molecules to biological function.

2

In Vivo NMR Spectroscopy – Techniques; Direct Detection; MRS; Kinetics and Labels; Fluxes; Concentrations

Robin de Graaf

Department of Diagnostic Radiology, Yale University School of Medicine, MR center, PO Box 208043, New Haven, CT 06520-8043, USA

2.1. INTRODUCTION

Nuclear magnetic resonance (NMR) is based on the magnetic properties of nuclei. When placed in a (strong) external magnetic field, the nuclei can be observed by the absorption and emission of electromagnetic

Metabolomics by In Vivo NMR. Edited by R. G. Shulman and D. L. Rothman
 ISBN: 0-470-84719-0

radiation. Purcell *et al.* (1) at MIT, Cambridge, MA, and Bloch *et al.* (2) at Stanford, CA, simultaneously but independently discovered NMR in 1945. In 1952 Bloch and Purcell shared the Nobel Prize for Physics in recognition of their pioneering achievements (1–4). At this stage, NMR was purely an experiment for physicists to determine the nuclear magnetic moments of nuclei. NMR could only develop to one of the most versatile forms of spectroscopy after the discovery that nuclei within the same molecule absorb energy at different resonance frequencies. These so-called chemical shift effects, which are directly related to the chemical environment of the nuclei, were first observed in 1949 by Proctor and Yu (5), and independently by Dickinson (6).

During the first two decades, NMR spectra were recorded in a continuous wave mode in which either the magnetic field strength or the radio frequency was swept through the spectral area of interest, whilst keeping the other fixed. In 1966, NMR was revolutionized by Ernst and Anderson (7), who introduced pulsed NMR in combination with Fourier transformation, leading to greatly enhanced sensitivity and increased versatility (e.g. multi-pulse experiments). Pulsed or Fourier transform NMR is at the heart of all modern NMR experiments.

The induced energy level difference of nuclei in an external magnetic field is very small when compared with the thermal energy, so that the energy levels are almost equally populated. As a result the absorption of photons (electromagnetic quanta) is very low, making NMR a very insensitive technique when compared with other forms of spectroscopy. However, as a consequence of the minimal perturbation of the spins by the external magnetic field, NMR is also a noninvasive and nondestructive technique, ideally suited for *in vivo* measurements. In fact, by observing the water signal from his own finger, Bloch was the first to use NMR on a living system. Soon after the discovery of NMR, others showed the possibility of using NMR to study living objects. In 1950, Shaw and Elsken (8) used proton NMR to investigate the water content of vegetable material. Odebald and Lindstrom (9) obtained proton NMR signals from a number of mammalian preparations in 1955. Continued interest in defining and explaining the properties of water in biological tissues led to the promising report of Damadian in 1971 (10) that NMR properties (relaxation times) of malignant tumorous tissues differ significantly from normal tissue, suggesting that (proton) NMR may have diagnostic value. In the early 1970s, the first high-resolution NMR experiments on intact living systems were reported. Moon and Richards (11) used ^{31}P NMR on intact red blood cells and showed how the intracellular pH can be determined from chemical shift differences. In 1974, Hoult *et al.* (12) reported the first study of ^{31}P NMR on intact, excised rat hindleg skeletal muscle. Four years later, the first ^{31}P NMR studies on excised and *in vivo* mammalian brain were performed by Chance *et al.* (13) and improved with surface coil detection by Ackerman *et al.* (14) in 1980. Technical difficulties, like water suppression, delayed the application of ^{1}H MRS to study brain metabolism. The first *in vivo* ^{1}H MRS spectrum from rat brain was obtained by Behar *et al.* (15) in 1983 at Yale University. Two years earlier the same group also performed the first *in vivo* ^{13}C MRS study on rat brain (16).

Around the same time that reports on *in vivo* NMR spectroscopy appeared, Lauterbur (17) and somewhat later Mansfield (18) described the first reports on a major constituent of modern NMR, namely *in vivo* NMR imaging or magnetic resonance imaging (MRI). By applying position-dependent magnetic fields in addition to the static magnetic field, they were able to reconstruct the spatial distribution of the spins in the form of an image. MRI has revolutionized the field of medicine and, in recognition of this achievement, Lauterbur and Mansfield shared the 2003 Nobel Prize. *In vivo* NMR spectroscopy or magnetic resonance spectroscopy (MRS) and MRI have evolved from relatively simple one- or two-RF pulse sequences to complex techniques involving spatial localization, water and lipid suppression and spectral editing for MRS and time-varying magnetic field gradients and ultrafast and multiparametric acquisition schemes for MRI.

Apart from MRI and MRS, there are many other tools available to study the anatomy, dynamics and metabolism of intact living tissues. For example, X-ray and computed tomography (CT) can provide high-resolution structural images, whereas others, including positron emission tomography (PET) and single photon emission computed tomography (SPECT) give rise to relatively low-resolution functional images. MRI has been shown to provide both high-resolution, high-contrast morphological images and high-resolution functional data. Furthermore, NMR does not require the use of ionizing radiation, making MRI a completely noninvasive and nondestructive imaging modality. The major advantage of *in vivo* NMR spectroscopy over the other mentioned modalities is, besides the noninvasive character, the excellent chemical specificity as expressed in the chemical shift and J coupling constants. This allows the study of specific metabolites and metabolic pathways, thereby offering a unique tool to study *in vivo* metabolism.

This chapter will give a brief introduction to *in vivo* NMR spectroscopy with an emphasis on spectral appearance and absolute quantification. It does not represent a complete review, for which the interested reader is referred to the literature (19). It is assumed that the reader is familiar with the basic concepts of (organic) NMR spectroscopy.

2.2. *IN VIVO* NMR SPECTROSCOPY

2.2.1. Proton NMR

The proton nucleus is, besides the low-abundance hydrogen isotope tritium, the most sensitive nucleus for NMR, both in terms of intrinsic NMR sensitivity (high gyromagnetic ratio) and high natural abundance (>99.9 %). Since nearly all metabolites contain protons, *in vivo* ^{1}H NMR is, in principle, a powerful technique to observe, identify and quantify a large number of biologically important metabolites in intact tissue.

However, while the application of ^{1}H NMR spectroscopy to study cerebral metabolism has greatly increased in the last decade, studies on skeletal muscle and liver have been limited, mainly because of technical reasons and limited information content. ^{1}H NMR spectra are always dominated by a large resonance from water. Most of the available water suppression techniques depend on high magnetic field homogeneity (and hence a narrow water resonance line). Although this is a general problem, high magnetic homogeneity is especially difficult to satisfy in the liver, due to the high iron deposits. Owing to the limited spatial resolution of NMR spectroscopy, tissue heterogeneity is always a consideration. However, it is especially important for muscle studies, since metabolic content significantly differs between muscle types. While ^{1}H NMR spectra of brain hold many resonances, the spectra from muscle and especially liver are relatively simple and are typically dominated by intense lipid resonances. Nevertheless, ^{1}H NMR on muscle and liver can provide important information on metabolism and a few selected examples will be presented.

2.2.1.1. Identification of resonances

The sensitivity of *in vivo* proton NMR typically limits the detection to compounds with a concentration higher than ~0.5 mM. Despite this inherent limitation, an *in vivo* ^{1}H NMR spectrum still holds many resonances from a wide range of different metabolites. Figure 2.1 shows a typical *in vivo* (water-suppressed) ^{1}H NMR spectrum obtained from human muscle at 4 T. The largest resonances typically originate from intramyocellular (IMCL) and extramyocellular (EMCL) lipids between 1.0 and 2.5 ppm. The major metabolite resonances are from trimethylammonium $N(CH_3)_3$ groups at ~3.22 ppm and the methylene CH_2 and methyl CH_3 groups of total creatine (tCr) at 3.91 and 3.03 ppm, respectively. Total creatine represents the sum of creatine and phosphocreatine. In the downfield region of the spectrum, two clear resonances of carnosine are visible. Other resonances that are present in ^{1}H NMR spectra of muscle are glutamate (2.34 ppm) and lactic acid (1.32 ppm), both of which are typically overwhelmed by resonances from IMCL and EMCL. Under ischemic conditions such as are encountered during exercise, the N-δ proton in the proximal F8 histidine of deoxymyoglobin can be observed.

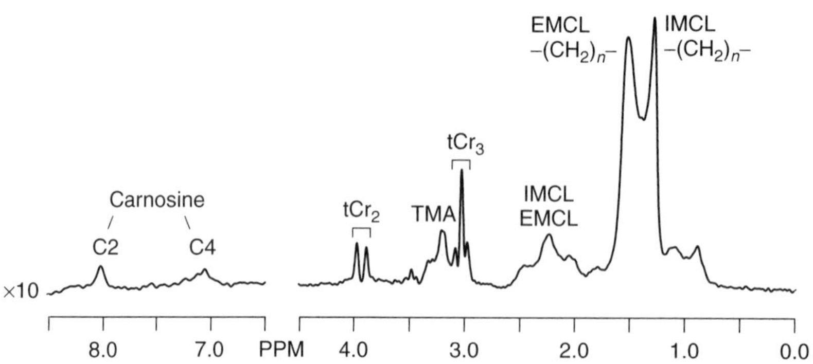

Figure 2.1. Proton NMR spectrum from human skeletal muscle *in vivo* [4000 μl, STEAM localization, $TR = 4000$ ms, $TE = 15$ ms, $TM = 15$ ms, number of experiments (NEX) = 128]. TMA, tetramethyl ammonium groups; tCr_2, total creatine methylene; and tCr_3, methyl groups.

A detailed discussion of the biological function and significance of the detected metabolites will be given in the following chapters. Here we briefly discuss several selected applications of ^{1}H NMR to study muscle physiology.

Carnosine and intracellular pH. Carnosine (β-alanyl-L-histidine), a dipeptide between alanine and histidine, is part of a series of compounds referred to as aminoacyl-histidine dipeptides. Other members include homocarnosine (γ-aminobutyryl-L-histidine) and anserine (β-alanyl-L-1-methyl-histidine). This group of naturally occurring histidine-containing molecules is particularly abundant in excitable tissues, such as muscle and nervous tissue. The biological functions of the aminoacyl-histidine dipeptides remain enigmatic, although the roles of pH buffer and antioxidant have been proposed. Carnosine was first observed over a century ago and has subsequently been found by a wide range of techniques, including ^{1}H NMR spectroscopy. The pK of the C-2 and C-4 protons on the imidazole ring of histidine are in the physiological pH range, providing a noninvasive method to measure intracellular pH with ^{1}H NMR spectroscopy *in vivo*. Figure 2.2 shows the pH dependence of the C-2 and C-4 proton chemical shifts of carnosine using published values from Pan *et al.* (20). The physical principle underlying the pH-dependent chemical shifts is discussed in Section 2.2.2.2 for ^{31}P NMR spectroscopy. Yoshizaki *et al.* (21) initially used carnosine to determine pH in excised frog muscle, which has subsequently been followed by *in vivo* studies,

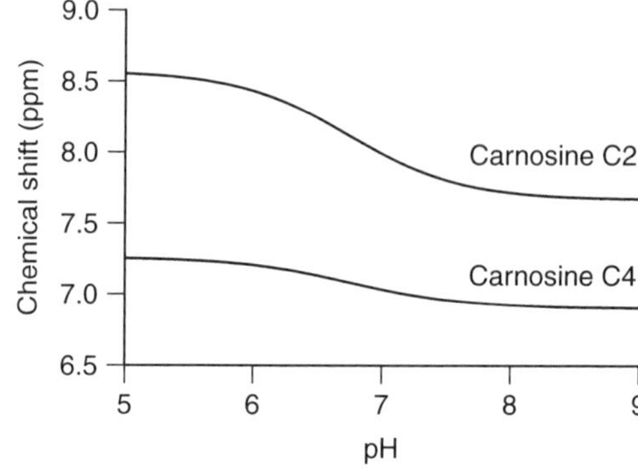

Figure 2.2. pH-dependence of the chemical shift positions of the C-2 and C-4 protons on the imidazole ring of histidine in carnosine.

including studies in human muscle. Several studies (20, 22) have show an excellent correlation between the determination of intracellular pH by ^{1}H (carnosine) and ^{31}P (inorganic phosphate) NMR.

Using carnosine for intracellular pH determination has several advantages over the more traditional methods using the chemical shift of inorganic phosphate in ^{31}P NMR spectra. First, the sensitivity of carnosine detection is very high due to the inherently high sensitivity of ^{1}H NMR, the high concentration of carnosine in skeletal muscle (up to 20 mM for human muscle) and the relatively short T_1 relaxation times. This is especially important in studies in which the concentration of inorganic phosphate significantly decreases (e.g. the recovery stage in muscle exercise studies). The high sensitivity of carnosine detection would allow a substantially improved time resolution in exercise studies. Second, the measurement of intracellular pH using carnosine is relatively insensitive to the presence of divalent cations, such as free magnesium (Mg^{2+}). This is especially important in exercise studies, where intracellular free Mg^{2+} concentrations increase.

Intramyocellular and extramyocellular lipids. ^{1}H NMR spectra of skeletal muscle are dominated by intense lipid resonances between 0.9 and 2.5 ppm. Schick *et al.* (23) and others (24–27) have assigned the resonances at *ca* 1.5 and 1.3 ppm to the methylene protons of EMCL and IMCL, respectively. The observed phenomenon of shifted resonances for identical chemical groups is the result of the geometrical arrangement of the lipids relative to the main magnetic field. Subcutaneous or interstitial adipose tissue (here summarized as EMCL) forms typically flat structures along the main axis of the muscle and extremity. IMCL, however, are stored in spherical droplets in the cytoplasm of muscle cells. The free or protein-bound lipids in the cytoplasm are of much lower concentration. Theoretical considerations based on bulk magnetic susceptibility frequency shifts for various geometrical orientations have been used to predict the frequency shifts of IMCL and EMCL. These frequency shifts can drastically change the appearance of ^{1}H NMR spectra as the angle of the muscle fibers changes relative to the external magnetic field. Similar effects are observed for creatine and carnosine (28).

While EMCL are metabolically relatively inert, there is evidence that, besides extramuscular energy sources, like glucose and free fatty acids, and intramuscular glycogen, IMCL are an energy storage form that is readily accessible during long-term exercise, particularly as they are primarily located immediately adjacent to mitochondria. Using ^{1}H NMR spectroscopy it has been found that IMCL levels are extremely sensitive to physical exercise and diet (29, 30). In diabetes several groups have observed an inverse correlation between IMCL levels and insulin sensitivity (31). It has also been shown that different muscle groups contain different levels of IMCL (26).

Deoxymyoglobin. Myoglobin (Mb), a 16.7 kDa protein, plays an important role in muscle physiology as an oxygen storage compound and a facilitator of oxygen diffusion. It has been shown by several groups that oxygen saturation in human skeletal muscle can be determined by detecting the deoxymyoglobin (DMb) signal by ^{1}H NMR spectroscopy at ~79 ppm (32–34). Despite its low concentration (~300 μM during muscle ischemia), the detection of DMb by ^{1}H NMR spectroscopy is possible because the resonance position of the N-δ proton in the proximal F8 histidine of DMb is sufficiently shifted downfield, away from the more intense resonances of water and lipids. The short T_1 relaxation time of ~10 ms allows substantial signal averaging and hence an improved sensitivity. The oxygenated form of Mb does not have any resonances with paramagnetic shifts and therefore is unobservable. Under normoxic conditions Mb is completely oxygenated, such that no signal from Mb can be detected at rest. However, under ischemic conditions, as achieved using a pressure cuff or during heavy exercise, a large DMb signal can be observed. Kreis *et al.* (34) have shown that the appearance and disappearance of DMb by the application and release of a pressure cuff can provide information on three separate processes: the total tissue Mb concentration is proportional to the maximum DMb (assuming complete deoxygentation): the basal rate of oxygen consumption is reflected in the rate of DMb accumulation during the induction of ischemia; finally, the exponential reoxygenation of DMb upon release of ischemia is determined by reperfusion constants, capillary density and the diffusion properties of DMb and oxygen in the muscle.

2.2.2. Phosphorus NMR

The success of *in vivo* proton NMR spectroscopy in routine (clinical) MR is only matched by phosphorus NMR. The relatively high sensitivity of phosphorus NMR (*ca* 7 % of protons), together with a 100 % natural abundance, allows the acquisition of high-quality spectra within minutes. Furthermore, the chemical shift dispersion of the phosphates found *in vivo* is relatively large (~30 ppm), resulting in excellent spectral resolution even at low (clinical) magnetic field strengths. Phosphorus NMR is very useful because with simple NMR methods it is capable of detecting all metabolites that play key roles in tissue energy metabolism. Furthermore, biologically relevant parameters such as intracellular pH may be indirectly deduced.

2.2.2.1. Identification of resonances

A typical *in vivo* phosphorus NMR spectrum holds a limited number of resonances (Figure 2.3). The exact chemical shift position of almost all resonances is sensitive to physiological parameters like intracellular pH and ionic (magnesium) strength. By convention, the phosphocreatine resonance is used as an internal chemical shift reference and has been assigned a chemical shift of 0.00 ppm. At a pH of 7.2, with full magnesium complexation, the resonances of adenosine triphosphate (ATP) appear at −7.52 (α), −16.26 (β) and −2.48 ppm (γ). The resonance of inorganic phosphate appears at 5.02 ppm. Under favorable, high-sensitivity conditions phosphorus NMR spectra can also hold resonances from phospho-monoesters and -diesters. Table 2.1 summarizes the chemical shift of the most commonly observed ^{31}P-containing metabolites. Note that phosphocreatine is completely absent in ^{31}P NMR spectra from liver.

2.2.2.2. Intracellular pH

The chemical shift of many phosphorus-containing compounds is dependent on a number of physiological parameters, in particular intracellular pH and magnesium concentration. The cause of this phenomenon can be found in the fact that protonation (or complexation with magnesium) of a compound changes the chemical environment of nearby nuclei and hence changes the chemical shift of those nuclei. When the chemical exchange between the protonated and unprotonated forms is slow, the two forms will have two separate resonance frequencies, with the resonance amplitudes indicating the relative amounts of the two forms. However, for most compounds observed with phosphorus NMR, the chemical exchange is fast relative to the NMR time scale and only a single, average resonance is observed. The resonance frequency

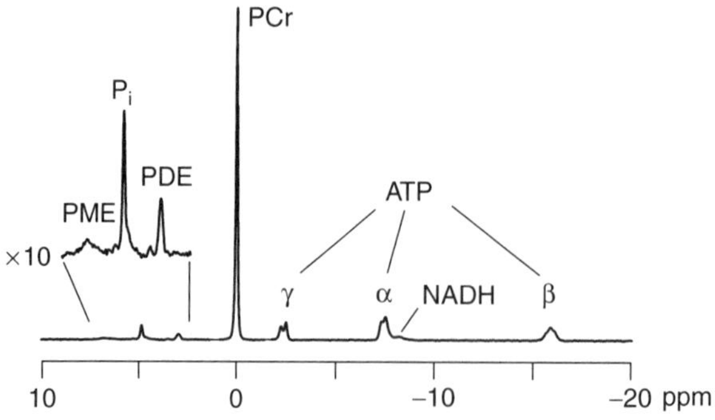

Figure 2.3. Phosphorus NMR spectrum from human muscle *in vivo* (surface coil localization, $TR = 10\,000$ ms, NEX = 32). P_i, inorganic phosphate; NADH, nicotinamide adenine dinucleotide; PDE, phospho-diesters; and PME, phospho-monoesters.

Table 2.1. Phosphorus chemical shifts for low molecular weight metabolites[a]

Compound		^{31}P chemical shift (ppm)
Adenosine monophosphate (AMP)		6.33
Adenosine diphosphate (ADP)	α	−7.05
	β	−3.09
Adenosine triphosphate (ATP)	α	−7.52
	β	−16.26
	γ	−2.48
Dihydroxyacetone phosphate		7.56
Fructose-6-phosphate		6.64
Glucose-1-phosphate		5.15
Glucose-6-phosphate		7.20
Glycerol-1-phosphate		7.02
Glycerol-3-phosphorylcholine		2.76
Glycerol-3-phosphorylethanolamino		3.20
Inorganic phosphate		5.02
Nicotinamide adenine dinucleotide (NADH)		−8.30
Phosphocreatine		0.00
Phosphoenolpyruvate		2.06
Phosphorylcholine		5.88

[a] Chemical shifts are reported relative to phosphocreatine at 0.00 ppm at pH 7.2 (fully complexed with magnesium).

is now indicative of the relative amounts of protonated and unprotonated form, and hence the pH can be described by a modified Henderson–Hasselbach relationship according to

$$\mathrm{pH} = \mathrm{p}K_{\mathrm{A}} + \log\left(\frac{\delta - \delta_{\mathrm{HA}}}{\delta_{\mathrm{A}} - \delta}\right) \tag{2.1}$$

where δ is the observed chemical shift, δ_{A} and δ_{HA} the chemical shifts of the unprotonated and protonated forms of compound A and pK_{A} the logarithm of the equilibrium constant for the acid–base equilibrium between HA and A.

Even though almost all resonances in ^{31}P NMR spectra have pH dependence, the resonance of inorganic phosphate is most commonly used for several reasons: its pK is in the physiological range (pK = 6.77); it is readily observed in most tissues (with muscle being a possible exception); and it has a large dependence on pH. Following similar arguments as outlined for intracellular pH, the free magnesium concentration can be deduced from the chemical shifts of ATP.

2.2.2.3. *Creatine kinase*

The application of magnetization transfer to phosphorus NMR has allowed the noninvasive measurement of the creatine kinase (CK) reaction

$$\mathrm{PCr}^{2-} + \mathrm{MgADP}^{-} + \mathrm{H}^{+} \longleftrightarrow \mathrm{MgATP}^{2-} + \mathrm{Cr} \tag{2.2}$$

The most generally accepted function of the CK/PCr (phosphocreatine) system is that of ‘temporal energy buffering’, i.e. during a high workload the levels of ATP remain constant through the conversion of PCr (which is present at relatively high concentrations) to ATP. Other, more controversial, functions include that of an 'energy transport system' and preventing the inactivation of ATPases by maintaining low intracellular ADP levels. The technique of magnetization transfer allows the determination of the absolute flux through the CK enzyme by following the fate of labeled (i.e. saturated or inverted) magnetization from either PCr or γ-ATP. The forward flux, from PCr to ATP, is a relatively straightforward measurement. However, the measurement of the reverse flux is complicated by ATP hydrolysis reactions by enzymes other than CK. Proper modifications of the magnetization transfer experiment (i.e. multiple saturations) do allow the correct measurement of the reverse flux (35). It is generally assumed that in skeletal muscle, heart and brain, the creatine kinase-catalyzed reactions are near or at equilibrium. Under this assumption, the intracellular ADP concentration can be calculated from Equation (2.2) if the equilibrium constant, intracellular metabolite (ATP, PCr, Cr) concentrations and pH are known.

2.2.3. Carbon-13 NMR

Phosphorus and proton NMR spectroscopy have been successfully employed in a range of (clinical) *in vivo* NMR studies. Nevertheless, both nuclei have inherent limitations. Phosphorus NMR spectra are normally characterized by a limited number of metabolites, while proton NMR is technically more challenging (e.g. water suppression) and suffers from inherently poor spectral resolution.

Carbon-13 NMR can offer complementary information to that obtained with phosphorus and proton NMR. Since (almost) all biologically relevant metabolites contain carbon, carbon-13 NMR is in principle capable of detecting many metabolites. Furthermore, the chemical shift dispersion extends over 200 ppm. However, the common carbon-12 isotope is not NMR-active. Carbon-13 does have a magnetic moment, but is only present at 1.1 % natural abundance. This, in combination with the low gyromagnetic ratio, makes carbon-13 NMR an inherently insensitive technique. Furthermore, strong heteronuclear scalar coupling interactions complicate the spectra and further reduce the sensitivity, such that double-resonance techniques (and hence additional, nonstandard hardware) must be employed to remove the effects of heteronuclear scalar coupling. Nevertheless, natural abundance carbon-13 NMR has been used to detect a wide range of metabolites. A typical natural abundance ^{13}C NMR spectrum from muscle is dominated by lipid resonances at 0–50, ~120 and ~180 ppm [Figure 2.4(a)]. However, ^{13}C NMR spectra also hold resonances from

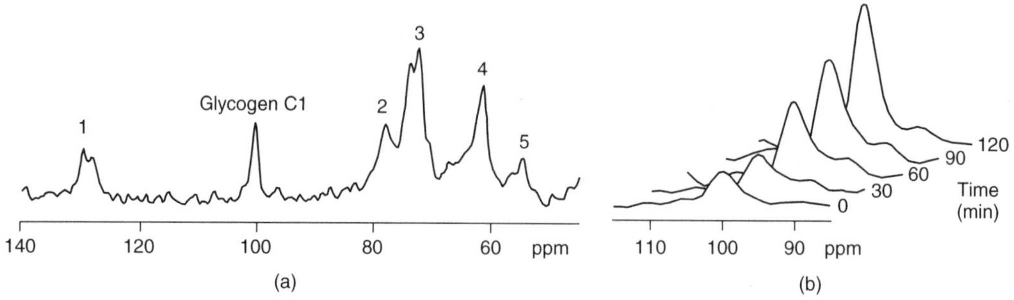

Figure 2.4. (a) Natural abundance carbon-13 NMR spectrum of rabbit liver *in vivo*. Peak assignments are for (**1**) olefinic carbons of fatty acyl chains (128.4 and 130.0 ppm), (**2**) glycogen-C4 (78.0 ppm), (**3**) glycogen-C3 (74.0 ppm), C2 and C5 (72.2 ppm) and C2 of glycerol backbone (69.5 ppm), (**4**) glycogen-C6 and C1/C3 of glycerol backbone and (**5**) trimethylammonium groups (e.g. choline, 54.6 ppm). The single resonance of glycogen-C1 is clearly visible at 100.5 ppm. (Reproduced from Gruetter R, Magnusson I, Katz LD, Shulman RG, Shulman GI, *Magn Reson Med* 1994; 31: 583–588 by permission of Wiley-Interscience.) (b) *In vivo* ^{13}C NMR spectra of human muscle glycogen-C1 (100.5 ppm) in a normal subject during a hyperglycemic–hyperinsulinemic clamp. The first spectrum represents the natural abundance glycogen-C1 resonance. (Reproduced from Shulman GI, Rothman DL, Jue T, Stein P, DeFronzo RA, Shulman RG, *New Engl J Med* 1990; **322**: 223–228 by permission of Massachusetts Medical Society.)

glycogen between 60 and 80 ppm, and especially [1-^{13}C]-glycogen at 100.5 ppm. The natural abundance signal of [1-^{13}C]-glycogen has been used to study carbohydrate metabolism during exercise (36) and fasting (37) and in diabetes (38). The low natural abundance of carbon-13 can be transformed into an advantage in that ^{13}C-enriched precursors can be infused to study metabolic pathways with little background interference from endogenous metabolites. Figure 2.4(b) shows an example of muscle glycogen turnover following intravenous infusion of [1-^{13}C]-glucose (39).

2.3. QUANTITATIVE NMR SPECTROSCOPY

NMR is, in principle, a quantitative technique and, as such, NMR spectra can be used to derive absolute concentrations of metabolites in animal and human tissues. This originates from the fact that the total integrated area under a resonance in a NMR spectrum (or the first data point of a free induction decay, FID) is proportional to the longitudinal thermal equilibrium magnetization vector, which in turn is directly related to the number of spins in the sample and hence to the concentration of that particular compound.

2.3.1. Experimental Considerations

In practice the derivation of absolute concentrations by NMR spectroscopy is not straightforward, since many additional factors can influence the metabolite resonance area. These factors can include T_1 and T_2 relaxation, diffusion, exchange, (partial) NMR invisibility, spectral overlap of resonances and the choice of an internal or external concentration reference. The literature on signal quantification by NMR is extensive [e.g. see de Graaf (19) for reviews], but can essentially be simplified to a three-step process. First, one has to ensure that the observed signal is directly proportional to the longitudinal thermal equilibrium magnetization M_0. Here the factors that can potentially affect the observed signal intensity will be summarized.

2.3.1.1. Longitudinal T_1 relaxation

When the repetition time, *TR*, of a single pulse sequence is shorter than four to five times the longitudinal relaxation time, T_1, the longitudinal magnetization cannot completely recover before the following excitation, eventually leading to a (lower) steady-state longitudinal magnetization given by

$$M_z(TR) = M_0 \frac{\left(1 - e^{-TR/T_1}\right)}{\left(1 - \cos\theta . e^{-TR/T_1}\right)} \tag{2.3}$$

where θ is the nutation angle. To obtain a measure of the thermal equilibrium magnetization, M_0, the acquired signal intensity of each resonance must be corrected for partial T_1 saturation. This can simply be achieved by using Equation (2.3) as a correction factor, but it requires knowledge of the T_1 relaxation time. Note that Equation (2.3) only holds for a single pulse-acquisition experiment. For more complicated experiments involving spin echo delays, Equation (2.3) needs to be modified to account for the additional radio frequency (RF) pulses, especially for short T_1 relaxation times and a long echo time, *TE*. With surface coils, correcting for partial saturation is further complicated since the nutation angle, θ, and consequently the saturation factor, depends on the position relative to the coil. This problem can be alleviated by executing the entire pulse sequence with adiabatic RF pulses. The correction for partial saturation can be omitted completely if the experiments are performed with $TR > 4T_{1\,max}$ ($T_{1\,max}$ being the longest T_1 relaxation time present), such that $M_z = M_0$ for all resonances. Even though this increases the experimental duration, the use of long repetition times is advisable since it eliminates systematic errors caused by application of an empirically determined T_1 saturation factor.

2.3.1.2. Transverse T_2 relaxation

In an experiment that utilizes spin- or stimulated-echo delays, *TE* signal losses are induced due to T_2 relaxation according to:

$$M_{xy}(TE) = M_{xy}(0).e^{-TE/T_2} \tag{2.4}$$

Often T_2 relaxation during a pulse sequence has been minimized by employing short echo times ($TE < 20\,\text{ms}$). However, hardware limitations like finite gradient ramp times or hardware imperfections like eddy currents may limit the shortest available echo time. Furthermore, for compounds such as ATP or macromolecules, even the shortest echo time leads to significant signal reduction. On the other hand, many groups have used long echo times ($TE > 100\,\text{ms}$) in order to reduce baseline oscillations, simplify the appearance of spectra and improve water suppression. In all cases a proper correction can only be made if the transverse relaxation time, T_2, is quantitatively known for each resonance.

2.3.1.3. Nuclear Overhauser effects

Related to longitudinal T_1 relaxation is the nuclear Overhauser effect. Under conditions of thermal equilibrium the relative populations of two spin states are given by the Boltzmann distribution. When this equilibrium distribution is disturbed, for instance during proton decoupling (saturation) in ^{13}C-[^{1}H] NMR, it can cause an increase in the equilibrium magnetization of the adjacent ^{13}C nucleus. The process of signal enhancement via a 'through-space' dipolar interaction between spatially adjacent nuclei is known as the nuclear Overhauser effect. For a ^{1}H–^{13}C interaction, the maximum enhancement is 2.988 for the ^{13}C nucleus. However, the maximum enhancement is only achieved when the molecular rotation ('tumbling') is very fast and if the spin–lattice relaxation of the observed nucleus is completely determined by the dipolar coupling. If either of these criteria is not met, the enhancement will be smaller and must typically be empirically determined. However, for the sake of simplicity and, more importantly, accuracy, it is better to design pulse sequences in such a manner that nuclear Overhauser effects are completely eliminated.

2.3.1.4. Diffusion

All pulse sequences employing spin or stimulated echo delays are affected by diffusion through microscopic susceptibility gradients in the tissue. The effects of diffusion are emphasized when magnetic field gradients are used. The effect of diffusion on absolute quantification can be pronounced when large molecules like ATP (with a low diffusion constant $D \approx 0.2 \times 10^{-3}\,\text{mm}^2/\text{s}$) are being calibrated against a low-molecular-weight compound like water ($D \approx 0.7 \times 10^{-3}\,\text{mm}^2/\text{s}$). In analogy to T_2 relaxation, this effect can be minimized by using shorter echo times. Furthermore, economical use of magnetic field gradients (e.g. small gradient duration, separation and amplitude) together with a well-shimmed sample and an optimized pulse sequence (e.g. Carr–Purcell–Meiboom–Gill echoes instead of Hahn echoes) can further minimize diffusion to a level where it does not significantly affect metabolite quantification.

It should be realized that these four parameters, and especially relaxation and diffusion, may change over time due to development of pathology or changes in temperature. For instance, in stroke the apparent diffusion coefficient of water decreases almost immediately after the onset of ischemia. In the more chronic phase of the ischemic lesion, the T_2 relaxation time of water significantly increases. It then becomes crucial to know these parameters quantitatively, or design the experiment as to minimize the effect of them.

2.3.1.5. Scalar coupling

Nuclei with magnetic moments influence each other either through space (dipolar coupling) or through chemical bounds (scalar coupling). In a liquid the dipolar interactions normally average out to zero due to rapid tumbling, so that no net interaction between nuclei remains. A notable exception is the dipolar splitting in ^{1}H NMR of muscle tissue. Even though there is no net interaction, the dipolar coupling does lead to relaxation. The interactions through chemical bonds do not average to zero, giving rise to splitting of resonances into several smaller lines. The resonances are separated by the scalar J coupling constant. The J coupling constant is independent of the applied external magnetic field, since it is based on the fundamental principle of spin–spin pairing. Typical magnitudes of J coupling constants are, ^{1}H–^{1}H (1–15 Hz), ^{1}H–^{13}C (100–200 Hz), ^{1}H–^{31}P (10–20 Hz) and ^{31}P–O–^{31}P (15–20 Hz). While scalar coupling adds a wealth of additional information on chemical structure to the NMR spectrum, it also complicates interpretation, especially in spin- and stimulated-echo experiments. While single resonances decay mono-exponentially by T_2 relaxation [Equation (2.4)], scalar coupled resonances have a complicated signal delay which is dependent on the J coupling constants, the exact spin system, applied RF pulses and the T_2 relaxation time constant, making determination of T_2 complicated, if not impossible. Several approaches can be followed to eliminate or compensate this effect. The use of short echo times ($TE < 20$ ms) reduces the amplitude and phase modulation of most scalar coupled metabolites. Another approach which can be followed is to choose TE as a multiple of $1/J$, such that the metabolite of interest is completely refocused. Unfortunately, unless the J coupling constants are similar, this approach only works for one metabolite (or one resonance of a metabolite). The amplitude and phase distortions caused by J-modulation can also be determined experimentally in model solutions. The knowledge obtained from the solutions can used to correct the observed signal *in vivo*. However, as with most other parameters it is better to minimize the effects of scalar coupling than to empirically determine them.

2.3.1.6. Localization

Since the absolute concentration is directly proportional to the localized volume, it is clear that the spatial localization needs to be accurate and identical for metabolites and reference compound. Especially for heteronuclear internal calibration (e.g. calibration of ^{31}P metabolites with the water signal), this is a challenging task. To reduce the duration of the experiment, ^{1}H MRS would preferably be performed with a single-scan technique as STEAM (stimulated echo acquisition mode) or PRESS (point-resolved spectroscopy). Owing to the short T_2 relaxation times, ^{31}P MRS is most often executed with ISIS localization. However, the actual localized volume of localization techniques can substantially differ and, even though these differences could be compensated, it is more convenient (and probably more accurate) to use the same localization technique for both experiments.

2.3.1.7. Frequency dependent amplitude and phase distortions

This effect can appear in many forms. An obvious example is given when binomial pulses are used for water suppression in ^{1}H MRS. The spectrum is amplitude-modulated according to $\sin^n \omega (n = 1, 2, 3, \ldots)$ where the higher-order binomial pulses also exhibit nonlinear phase distortions. Simulations based on the Bloch equations or empirical determination of these distortions typically achieve adequate compensation. Other areas where amplitude distortions can play a role are in ^{31}P and ^{13}C MRS, where the effective chemical shift range is much larger than for ^{1}H NMR spectroscopy. When the RF amplitude is small with respect to this chemical shift dispersion, the nutation angle becomes frequency-dependent. This can result in substantial errors when, for instance, the β-ATP resonance (which is normally on the edge of the *in vivo* ^{31}P chemical shift range) is used for quantification. The use of smaller nutation angles will reduce these effects. A less obvious example of frequency-dependent signal modulation arises when signal is observed in the presence of frequency-selective RF pulses during magnetic field gradients. The difference in chemical shift between the resonances leads to localized volumes at different spatial positions. This chemical shift artifact can become dominant for wide spectral bandwidths, as encountered in ^{31}P NMR. For scalar coupled spin systems an additional effect occurs in that the refocusing of scalar evolution will vary at different spatial positions. These effects are difficult to simulate or measure, but they can be minimized by using RF pulses with large bandwidths.

The application of adiabatic half-passage pulses is very popular in (nonlocalized) ^{31}P NMR spectroscopy (and to a lesser degree ^{13}C NMR spectroscopy). However, for adiabatic pulses the RF amplitude should also be large enough to excite the entire chemical shift range uniformly. When the (off-resonance) adiabatic condition is not satisfied, substantial errors will arise if these effects are not taken into account. Again, the effects of RF pulses can be exactly calculated from the Bloch equations such that complete compensation is possible.

2.3.1.8. NMR visibility

The line width of resonances is inversely proportional to the T_2 relaxation time, which is related to the rotational mobility of the metabolite. Metabolites with low mobility (e.g. bound to macromolecular structures) give rise to very short T_2 relaxation times and hence broad resonances which can be unobservable in conventional NMR spectra. This will in turn lead to an underestimation of the concentration. Furthermore, if a water suppression technique like presaturation is used, the narrow resonance line arising from the mobile component of the metabolite under investigation may decrease due to magnetization transfer effects, leading to a further underestimation of the true concentration.

Considering the factors described above, a generally applicable formula can be constructed for the concentration of a metabolite.

$$S_{\mathrm{m}} = S_{\mathrm{mm}} C_{T1,\mathrm{m}} C_{T2,\mathrm{m}} C_{\mathrm{nOe,m}} C_{\mathrm{ADC,m}} C_{J,\mathrm{m}} C_{\mathrm{loc,m}} C_{\mathrm{RF,m}} \tag{2.5}$$

$$S_{\mathrm{r}} = S_{\mathrm{rm}} C_{T1,\mathrm{r}} C_{T2,\mathrm{r}} C_{\mathrm{nOe,r}} C_{\mathrm{ADC,r}} C_{J,\mathrm{r}} C_{\mathrm{loc,r}} C_{\mathrm{RF,r}} \tag{2.6}$$

$$[m] = \left(\frac{S_{\mathrm{m}}}{S_{\mathrm{r}}}\right) [r] C_{\mathrm{n}} C_{\mathrm{av}} \tag{2.7}$$

where S_{mm} = measured metabolite signal; S_{m} = corrected metabolite signal; S_{rm} = measured reference signal; S_{r} = corrected reference signal; C_{T1} = correction factor for partial saturation due to incomplete T_1 relaxation; C_{T2} = correction factor for T_2 relaxation [Equation (2.4)]; C_{nOe} = correction factor for nuclear Overhauser effects; C_{ADC} = correction factor for diffusion, C_J = correction factor for amplitude and phase modulations due to J coupling evolution; this factor can include the effects of frequency selective RF pulses on scalar-coupled spins; C_{loc} = correction factor for deviations from the ideal localization profile; C_{RF} = correction factor for amplitude and phase distortions due to specific RF pulse combinations like binomial RF pulses; $[m]$ = concentration of the metabolite under investigation; $[r]$ = concentration of the reference compound; C_{n} = correction for the number of equivalent nuclei for each resonance; and C_{av} = correction for the number of averages.

The factor of partial NMR invisibility is difficult to correct for without invasive measurements, like biopsies (see also Section 2.3.5.4). When some factors do not affect the measured signal, the corresponding correction factor equals 1. Each individual calibration technique needs, besides the general applicable factors in Equations (2.5)–(2.7), its own specific corrections, which will be described next.

2.3.2. Internal Concentration Reference

The strategy of an internal concentration reference is straightforward. The (corrected) resonance areas in the acquired spectrum are compared with that of a stable endogenous reference compound. For ^{1}H MRS, water, total creatine and N-acetyl aspartate (NAA; for brain) have been proposed as endogenous concentration references. However, it should always be kept in mind that the concentration of endogenous concentration references can change, for example during development, under pathological conditions or between species. NAA shows a substantial decrease in a wide range of neurodegenerative pathologies ranging from ischemia to Alzheimer's disease. Total creatine and water are relatively stable metabolites, although changes in their rotational mobility may change their NMR visibility. Furthermore, the concentration is often a function of spatial position, e.g. different concentrations between gray and white matter or between muscle groups. When using water as an internal concentration reference, it is especially important to discriminate between water from different compartments, like cerebral tissue and cerebro-spinal fluid, which may differ by 30–40 % in water content. This discrimination can be achieved on the basis of a double-exponential decay of the water. Alternatively, the compartments could be retrieved from high-resolution MR images with appropriate T_1 contrast. Despite the potential difficulties, water has been used by several NMR groups with some degree of success, judging by the favorable comparison with concentrations obtained with other techniques. For ^{31}P MRS, ATP and water have been used as endogenous concentration markers. ATP is only a suitable reference compound in those applications where the system under investigation is only mildly challenged. With severe pathologies like ischemia, anaerobic glycolysis rapidly consumes the available phosphocreatine pool (through the creatine kinase equilibrium), after which the ATP resonances start to decline.

In applications where the peak ratios are used, the total phosphate pool, defined as

$$[\mathrm{P_{tot}}] = [\mathrm{PME}] + [\mathrm{P_i}] + [\mathrm{PDE}] + [\mathrm{PCr}] + 3[\mathrm{NTP}] + 2[\mathrm{NAD} + \mathrm{NADH}] \tag{2.8}$$

can be used. The total phosphate pool should in principle be constant (assuming that none of the metabolites become NMR-invisible and none are transported out of the cells), such that changes in peak ratio (e.g. $[\mathrm{PCr}]/[\mathrm{P_{tot}}]$) can be attributed to a single metabolite (i.e. [PCr]).

The water reference is in practice equally suitable for ^{1}H and ^{31}P NMR spectroscopy. For internal water referencing, Equation (2.5) should be modified and extended to

$$[m] = \left(\frac{S_\mathrm{m}}{S_\mathrm{water}}\right)[\mathrm{water}]C_\mathrm{n}C_\mathrm{av}C_\mathrm{wc}C_\mathrm{HX} \tag{2.9}$$

where S_water = corrected water signal; [water] = water concentration (110 mol/l); C_wc = correction for the water content in the VOI – C_wc equals ~0.82 for gray matter, ~0.73 for white matter, >0.95 for cerebrospinal fluid (CSF) and ~0.78 for skeletal muscle; and C_HX = correction factor for the relative sensitivities between the proton and X channels (X = ^{31}P or ^{13}C).

The mentioned factors, C_wc, for different tissue types are from biochemical measurements. It should be realized that this factor will be reduced if NMR-invisible pools are present. The factor C_HP (i.e. X = ^{31}P) can be assessed by performing a phantom experiment with a phosphorus metabolite of known concentration in water. C_HP is then calculated by dividing the ^{1}H signal per mol/l of protons by the ^{31}P signal per mol/l of phosphorus. Since C_HP depends on a number of factors, including coil load, it is advisable to mimic the *in vivo* conditions as close as possible. For the homonuclear calibration strategy C_HH obviously equals 1.

2.3.3. External Concentration Reference

The method that utilizes an external concentration reference can be executed as shown in Figure 2.5(b). After collection of the desired *in vivo* spectrum, a reference spectrum from a calibration sample is obtained. To minimize the effects of RF inhomogeneities, the two voxels are chosen symmetrically about the center

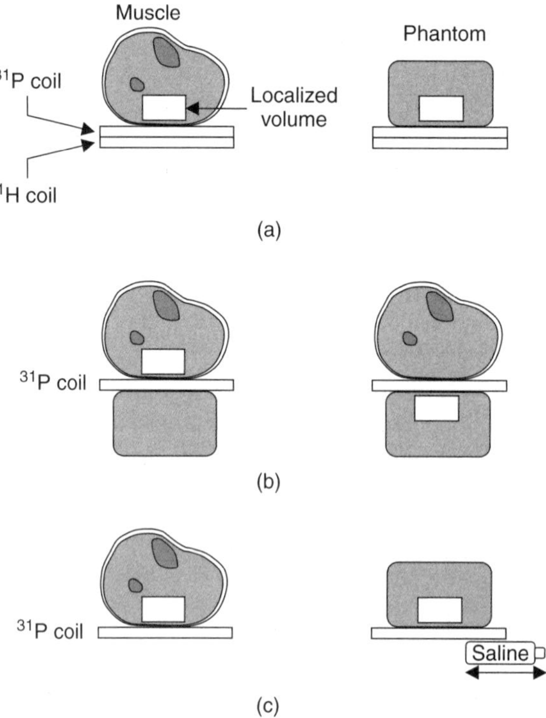

Figure 2.5. Calibration strategies for the quantification of metabolite concentration in human skeletal muscle for ^{31}P NMR spectroscopy. The left-hand set indicate the acquisition of metabolite spectra, while the right-hand set indicate the acquisition of the reference compound spectra. (a) Internal concentration reference. When water is used as an internal concentration reference, a separate *in vitro* reference acquisition is required to establish the relative ^{1}H and ^{31}P sensitivities. When a ^{31}P-containing compound is used, only the *in vivo* acquisition is required. (b) External concentration reference. (c) External simulated phantom concentration reference. A small saline-filled bottle is inserted in (or retracted from) the coil, in order to equalize the *in vivo* and *in vitro* coil loads.

of the coil. Alternatively, the RF inhomogeneity can be accounted for by simulation of the RF field distribution of the particular RF coil used. The metabolite concentration can be calculated according to:

$$[m] = \left(\frac{S_{\mathrm{m}}}{S_{\mathrm{r}}}\right)[r]C_{\mathrm{n}}C_{\mathrm{av}} \tag{2.10}$$

where S_{r} = corrected reference signal; and $[r]$ = concentration of external reference.

Equation (2.10) immediately shows the main distinction with the method of internal water calibration, in that no assumption needs to be made for the internal water concentration (which may vary with pathology, age and voxel composition). Furthermore, for heteronuclear experiments no calibration factor for the relative ^{1}H and ^{31}P (or ^{13}C) sensitivities is required.

The method of external concentration referencing is relatively simple since the experimental setup or the position of the patient need not be changed. Furthermore, it is a robust method, mainly being hampered by a dependency on the B_1 field distribution (which can be minimized by symmetrical placement of the sample with respect to the coil). The method can also be very time-efficient if the two localized volumes are acquired simultaneously with Hadamard or two-volume ISIS (or STEAM/PRESS) localization.

2.3.4. External Simulated Phantom Concentration Reference

Another method of quantification is aimed at simulating (human) tissue as closely as possible with a phantom of known composition [Figure 2.5(c)]. Because almost all systematic errors like RF inhomogeneity and localization affect the tissue and phantom in identical ways, this calibration method is in principle very robust. The method is only complicated by differences in coil loading between the tissue and the phantom. Two methods are available to compensate for differences in coil loading, i.e. load adjustment and load correction. For load adjustment, the electrical conductivity of the solution in the phantom is slightly lower than that of human tissue ($s \approx 0.64$ S/m) such that it allows for fine adjustment of the coil load with a second (smaller) phantom containing, for example, saline. During the procedure of load adjustment, the matching capacitance of the RF coil is left unchanged at the end of the *in vivo* experiment. After removal of the patient and accurate positioning of the phantom, the matching is optimized by slowly inserting the saline bottle (i.e. increasing the coil load). When the *in vitro* matching equals the previous *in vivo* matching, the *in vivo* and *in vitro* coil loads are identical. The concentration can then simply be calculated with Equation (2.5).

The method of load correction involves the addition of an external capillary which is measured nonselectively (nonlocalized) during the *in vivo* and *in vitro* experiments. Most conveniently, a compound is used which falls outside the spectral region of interest [e.g. tetramethylsilane (TMS) for ^{1}H or ^{13}C MRS and phenylphosphonicacid (PPA) or hexamethylphosphorustriamide (HMPT) for ^{31}P MRS]. The correction factor for difference in coil load is then calculated from the capillary signals obtained from the *in vivo* and *in vitro* experiments according to:

$$C_{\text{load}} = \frac{S_{invitro}}{S_{invivo}} \tag{2.11}$$

after which the concentration can be calculated as

$$[m] = \left(\frac{S_{\text{m}}}{S_{\text{r}}}\right)[r]C_{\text{n}}C_{\text{av}}C_{\text{load}} \tag{2.12}$$

Especially at high magnetic field strength, the method of external simulated phantom concentration reference may also be complicated by different B_1 magnetic field distributions *in vivo* and *in vitro*. This effect is difficult to correct, making this method less desirable at high magnetic fields (>3 T).

Although there is no universal calibration technique that is optimal under all conditions, it can be stated that homonuclear internal concentration calibration is the simplest to implement and provides good results for normal or mildly pathological conditions. Heteronuclear concentration calibration using internal water is more difficult to implement since it requires two RF coils and more correction factors. Generally, it was found to be less robust than methods using external concentration calibration. Under severe pathological conditions, internal concentration referencing is generally not an option. The two external phantom-based methods are relatively easy to implement and provide a similar accuracy of metabolite quantification.

Following acquisition of the experimental data (together with the required correction factors), the spectra need to be analyzed in terms of resonance areas. Several methods are available to accomplish resonance area quantification.

2.3.5. Signal Quantification

2.3.5.1. General principles

Assuming that the observed signal is directly proportional to the longitudinal equilibrium magnetization, M_0 (or that appropriate correction factors have been obtained), the free induction decay of a single frequency is given by

$$M_{xy}(t) = M_0 \mathrm{e}^{-i\Delta\omega t}.\mathrm{e}^{-i[\phi_0+\phi_1]}.\mathrm{e}^{-t/T_2} \tag{2.13}$$

where $\Delta\omega$ is the frequency offset ω relative to the Larmor frequency ω_0 (i.e. $\Delta\omega = \omega - \omega_0$). ϕ_0 and ϕ_1 represent the zero (constant) and first (linear) order phases, while T_2 is the transverse relaxation time constant. In the case that the start of acquisition is delayed by an amount Δt, the linear phase ϕ_1 is given by $(\omega - \omega_0) \cdot \Delta t$ (i.e. there will be a linear phase distribution across the NMR spectrum). The linear phase contribution can always be eliminated by proper experimental design, such that it can be ignored in further discussions.

It follows that the NMR signal is completely described by four parameters, namely M_0, T_2, $\Delta\omega$ and ϕ_0. For quantification purposes only, the longitudinal magnetization, M_0, is relevant. While the NMR signal is observed in the time-domain [Equation (2.13)], it is typically processed by Fourier transformation and displayed in the frequency domain. The observed frequency-domain NMR spectrum signal S is given by:

$$S(\omega) = A(\omega)\cos\phi_0 - D(\omega)\sin\phi_0 \tag{2.14}$$

$$A(\omega) = \frac{M_0 T_2}{1 + (\omega - \omega_0)^2 T_2^2} \quad \text{and} \quad D(\omega) = \frac{M_0 T_2^2(\omega - \omega_0)}{1 + (\omega - \omega_0)^2 T_2^2} \tag{2.15}$$

The frequency-domain signal $S(\omega)$ is in general a mixture of absorptive $A(\omega)$ and dispersive $D(\omega)$ line shapes, since the zero-order phase, ϕ_0, is typically not zero. However, dispersive line shapes are not desirable since they are broad and have no net integrated signal intensity (Figure 2.6). The process of phase correction can eliminate the dispersive component by making $\phi_0 = 0$, leaving only the absorptive component (Figure 2.6). Although Equations (2.13)–(2.15) appear very different, it is important to realize that the time and frequency representations of the NMR signal are equivalent, being related by a linear (and reversible) Fourier transformation. Table 2.2 summarizes the relations between the time and frequency-domain parameters describing a Lorentzian line shape, while Figure 2.6 displays a number of simulated examples. It follows that the T_2 relaxation time constant is inversely proportional to the line width at half maximum, $\Delta\nu_{1/2}$, according to:

$$\Delta\nu_{1/2} = \frac{1}{\pi T_2} \tag{2.16}$$

The objective of absolute signal quantification in NMR spectroscopy is to obtain a reliable estimate of M_0, which is directly proportional to the absolute concentration. For a time-domain signal (FID) holding a single frequency, M_0 is directly proportional to the amplitude of the first data point. However, for multiple frequencies, the FID and hence the first data point will be a summation of the amplitudes of all frequencies. In the frequency domain M_0 is proportional to the integral of the absorption resonance signal [Equation (2.15)]. The peak height is proportional to $M_0 T_2$ and does not therefore reflect the absolute concentration. However, in analogy to the time domain, the estimation of M_0 may become difficult in the frequency domain when multiple, partially overlapping resonances are present. In the next section several options of time- and frequency-domain resonance quantification will be discussed.

2.3.5.2. Integration

Integration of resonances is the most straightforward method of achieving spectral quantification. In high-resolution, liquid-state NMR this is normally an adequate method since resonances are well separated without any baseline fluctuations. Nevertheless, care should be taken to avoid systematic errors that may arise from incorrect integration boundaries or baseline offsets. The integration of *in vivo* NMR spectra poses a number of additional problems. Since most *in vivo* NMR spectroscopy experiments are performed at

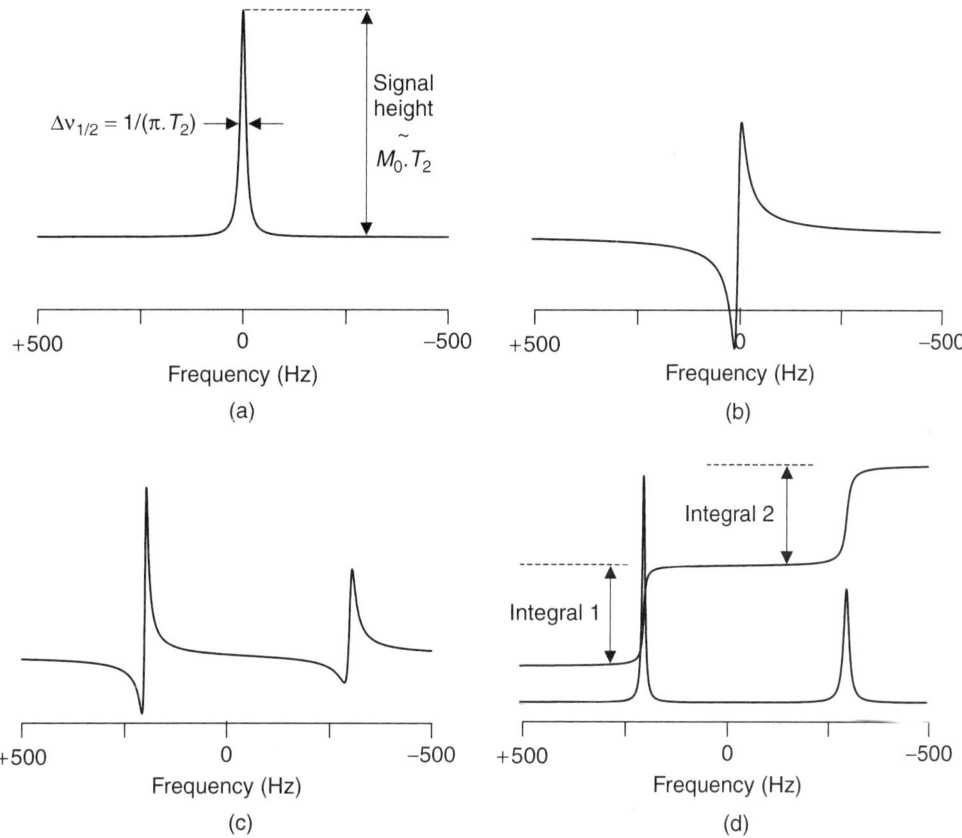

Figure 2.6. Principal components of a NMR spectrum. Complex Fourier transformation of an exponentially decaying FID (time-domain signal) gives rise to Lorentzian (a) absorption and (b) dispersion line shapes. Note that an absorption line is much narrower than a dispersion line. The frequency width at half maximum of the absorption line shape $\Delta\nu_{1/2}$ is inversely proportional to the (apparent) T_2 relaxation time constant. (c) In general, the initial phase ϕ_0 of an FID is nonzero, such that a mixture of absorption and dispersion line shapes is obtained. The dispersive component can be eliminated by 'phasing' the spectrum, such that only the absorptive component remains, as shown in (d). Note that the longitudinal equilibrium magnetization, M_0, is equal for both resonances. However, since the T_2 relaxation time constants are different, the peak heights are not equal. However, integration reveals that the number of nuclei (i.e. M_0) giving rise to the resonances is identical, since integral 1 = integral 2.

Table 2.2. Relationships between parameters in the time and frequency domain assuming an exponentially damped, sinusoidal time-domain signal

Time domain	Frequency domain
Frequency, ν ($= \omega/2\pi$)	Resonance position ν on the frequency axis
Relaxation rate, R_2 ($= 1/T_2$)	Linewidth (i.e. full width at half maximum) $\Delta\nu = R_2/\pi$ [$= 1/(\pi T_2)$]
Amplitude, M_0 (first point)	Total integrated area under the absorption part of the Lorentzian line ($-\infty < \nu < +\infty$)
Phase, ϕ	Phase ϕ of the Lorentzian line

relatively low magnetic field strengths, spectral overlap is the rule. In case of partial overlap of resonances, the integration boundaries can be selected to only integrate the resonance area that is not overlapping. However, the choice of integration boundaries will typically be operator-dependent, such that a rigorous method for selection must be developed in order to avoid systematic errors. In the case of complete spectral overlap of resonances, integration (or any other method) cannot be used for quantification without additional information, like nonoverlapping resonances of the compounds under investigation in another part of the NMR spectrum. A more severe problem of integration is encountered when broad resonances are present in the NMR spectrum. Immobilized molecules have a short T_2 relaxation time constant and give rise to broad resonances extending throughout large parts of the NMR spectrum. In these cases, integration must be combined with a baseline correction algorithm in order to remove the baseline and obtain an accurate estimate of the resonance intensities.

2.3.5.3. Least-squares fitting

Integration with baseline correction is a standard utility on most MR systems and under certain conditions (no severe baseline distortion or spectral overlap) it can provide reasonable estimates of resonance areas and hence of absolute concentrations. However, under many conditions, like ^{1}H NMR spectroscopy, integration is not an adequate quantification method and other options need to be explored. The large group of methods employing least-square fitting is rapidly becoming the method of choice for spectral quantification. The general principle of these methods is that the theoretical (fitted) spectrum should resemble the experimental spectrum as closely as possible by varying the parameters M_0, T_2, $\Delta\omega$ and ϕ_0 for each resonance. After many years of development, two methods appear to have crystallized as being best suited to *in vivo* NMR spectroscopy. VARPRO (40) and related techniques appear to be well suited to fitting of relatively few resonances, like those observed in ^{31}P NMR spectra. This time-domain algorithm is flexible for varying chemical shifts and scalar couplings, can readily avoid problems associated with intense baselines and is now routinely used (41, 42). LCmodel (43–45) is becoming the method of choice for more complex resonances, like those observed in short-echo-time ^{1}H NMR spectra. The LCmodel algorithm approximates the *in vivo* NMR spectrum as a linear combination of *in vitro* model solution spectra. Since all resonances of a given compound, e.g. glutamate, are fitted simultaneously, the great complexity of the glutamate NMR spectrum actually helps the convergence.

2.3.5.4. Example: glycogen

Carbohydrate reserves are mainly stored as glycogen in animals and humans. It is particularly abundant in muscle and liver, reaching concentrations up to 30–100 and 100–500 mmol/kg, respectively. Glycogen is also present in the brain, residing in astroglia at a concentration of *ca* 5 mmol/kg. The regulation of glycogen synthesis and breakdown plays an important role in systemic glucose metabolism and is crucial in the understanding of diseases such as diabetes mellitus.

The primary structure of glycogen consists of α-[1,4]-linked glucose chains containing 12–13 glucose residues. The chains are linked together through α-[1,6] branch points to form larger, spherical units referred to as β-particles. The β-particles can be further arranged into larger rosette-type structures termed α-particles. Despite its high molecular weight (10^7–10^9 Da), glycogen gives rise to narrow ^{1}H and ^{13}C NMR resonances both *in vitro* and *in vivo*, indicating a high degree of internal mobility. Sillerud and Shulman (46) reported in 1983 the surprising result that the [1-^{13}C]-glycogen resonance at 100.5 ppm is ~100 % visible. This careful investigation on the NMR visibility of glycogen set the stage for subsequent studies where these early results have been extended and confirmed. Visibility was established *in vitro* from different glycogen samples and *in situ* from perfused rat liver, where the intensities of glycogen were compared with those from glucose obtained by glycogen degradation using glucagon or amyloglucosidase.

The percentage yield was 98 % when averaged over all data, with no difference between *in vitro* and *in situ* results. Furthermore, since extractions were made at different degrees of hydrolysis without differences in visibility, there was no reason to believe that the visibility depended upon the molecular weight of glycogen.

Studies on the ^{1}H and ^{13}C relaxation properties of glycogen support the high NMR visibility of glycogen. Zang *et al.* (47) found that the ^{13}C T_1 relaxation of extracted rabbit liver glycogen can be described by a rotational correlation time τ_c of 4–6 ns, while Chen *et al.* (48) showed that the ^{1}H T_1 relaxation can be described with $\tau_c = 2.7$ ns. The original observation by Sillerud and Shulman gave an average value of $\tau_c = 4.6$ ns. These studies indicated that the correlation times which dominate the ^{13}C and ^{1}H dipolar interactions are much shorter than the molecular rotation correlation time [τ_c(molecular) $\approx$10 μs]. However, despite the high internal mobility of glycogen and the repeated detection by ^{13}C NMR spectroscopy, the NMR visibility of glycogen remains an issue of debate (49). Here we briefly review the process of glycogen quantification.

As discussed in Section 2.3.5.1, the spectroscopic line width of a resonance is inversely proportional to the (apparent) T_2 relaxation time constant. The theory of dipolar relaxation predicts progressively shorter T_2 relaxation times with increasing rotation correlation times and hence with increasing molecular size. When the T_2 relaxation time becomes shorter than a few milliseconds, the resonance line will have a width of hundreds of Hertz and will essentially be merged with the spectroscopic baseline, leading to an underestimation of the actual concentration. When the T_2 relaxation time is on the order of 10 ms, the spectroscopic line can be observed, but may be (partially) overlapping with other resonances, especially at low magnetic fields. The exact integration boundaries as well as the contribution of other resonances become crucial to establish the correct concentration. Only when the T_2 relaxation time constant is larger than *ca* 10 ms will the resonance line be sufficiently narrow to allow proper integration. The observation of narrow glycogen resonances indicates that at least part of the glycogen has a high rotational mobility and hence a longer T_2 relaxation time. However, this does not exclude the possibility that glycogen has multi-exponential T_2 relaxation characteristics. Overloop *et al.* (50, 51) showed that hepatic glycogen in solution has multi-exponential T_2 relaxation, with a continuous T_2 distribution from 1 to 30 ms. Furthermore, the T_2 relaxation characteristics of glycogen changed for different molecular configurations such as β-particles and larger aggregated α-particles. However, Overloop *et al.* (50) demonstrated that, with carefully selected integration boundaries or multiexponential fitting, 100 % of the [1-^{13}C]-glycogen resonance area can be obtained. This leads to the conclusion that multiexponential T_2 relaxation does not decrease the NMR visibility of glycogen.

Another complication in the quantification of glycogen could be potential overlap with other ^{13}C NMR resonances (49). Despite the large chemical shift dispersion for ^{13}C NMR of >200 ppm, spectral overlap of resonances can still be a problem, especially at lower magnetic field strengths. At 2.1 T, but even at 1.5 T (52) the [1-^{13}C]-glycogen signal resonates in the middle of a broad, flat baseline without other signals (Figure 2.4), allowing accurate quantification within the signal-to-noise ratio.

Muscle glycogen *in vivo* was also shown to be ~100 % NMR visible by comparing intensities of rabbit muscle [1-^{13}C]-glycogen *in vivo* with an *in vitro* glycogen control sample (53) and by comparing [1-^{13}C]-glycogen resonance intensities from human gastrocnemius muscle with biopsy samples from the same muscles (54). These results, shown in Figure 2.7, also emphasize the greater accuracy of glycogen ^{13}C NMR spectroscopy vs biopsy, as shown by the relative error bars. In the liver *in vivo* Shalwitz *et al.* (55) also observed ~100 % NMR visibility of [1-^{13}C]-glycogen. With this validation of the ^{13}C NMR method, many noninvasive results have been obtained from human *in vivo* measurements of hepatic glycogen (56). While the majority of studies support ~100 % NMR visibility of glycogen *in vivo* (46, 52–54, 57) and *in vitro* (58), there are several studies that report a lower visibility (59, 60). However, most of the reports on reduced NMR visibility have not performed a direct comparison between the NMR measurement and an invasive but established method, like biopsies or the detection of the degradation products of glycogen.

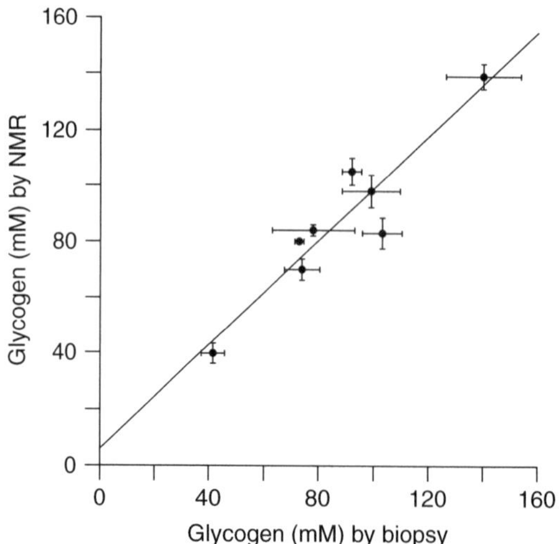

Figure 2.7. Correlation between measurements of muscle glycogen concentration by NMR and biopsy. Vertical and horizontal error bars show the standard deviation for repeated NMR ($n = 6$) and biopsy ($n = 3$) measurements on each of the eight subjects, respectively. The major contributor of the variation in NMR results is the signal-to-noise ratio of ~20:1. The contributors to errors in biopsy results are sampling within a nonhomogeneous tissue, loss of tissue during homogenization and possible metabolism of glycogen after biopsy and before freeze-clamping. (Reproduced from Taylor R, Price TB, Rothman DL, Shulman RG, Shulman GI, *Magn Reson Med* 1992; **27**: 13–20 by permission of Wiley-Interscience.)

In one case, a subsequent report from the same laboratory acknowledged that the glycogen NMR visibility was ~100 % (52). With this in mind, most of the reservations about the NMR visibility are deemed unsubstantiated and any future claims should be accompanied by carefully acquired experimental data.

Proton NMR spectra of glycogen in deuteriumoxide also showed well-resolved ^{1}H resonances, which were approximately 100 % NMR visible when compared with the hydrolyzed glucose moieties. The ^{1}H-glycogen resonance at 5.38 ppm was especially well separated (61). Magnetization transfer between the glycogen proton resonances and water was observed to dominate relaxation in aqueous solutions. It was proposed that the magnetization transfer is mediated by the exchange of a nearby hydroxyl proton (62). As a result, water suppression (or any other perturbation of the water) to detect the proton resonances of glycogen is counter-productive, since the magnetization transfer process significantly decreases the glycogen proton intensities. Although small resonances from ^{1}H-glycogen have been observed in the rat liver *in vivo* (63), the reduced intensity has not allowed proton NMR to yield important results *in vivo*.

REFERENCES

1. Purcell EM, Torrey HC, Pound RV, Resonance absorption by nuclear magnetic moments in a solid. *Phys Rev* 1946; **69**: 37–38.
2. Bloch F, Hansen WW, Packard ME, Nuclear induction. *Phys Rev* 1946; **69**: 127.
3. Bloch F, Nuclear induction. *Phys Rev* 1946; **70**: 460–473.
4. Bloch F, Hansen WW, Packard ME, The nuclear induction experiment. *Phys Rev* 1946; **70**: 474–485.
5. Proctor WG, Yu FC, The dependence of a nuclear magnetic resonance frequency upon chemical compound. *Phys Rev* 1950; **77**.

6. Dickinson WC, Dependence of the F19 nuclear resonance position on chemical compound. *Phys Rev* 1950; **77**: 736.
7. Ernst RR, Anderson WA, Applications of Fourier transform spectroscopy to magnetic resonance. *Rev Sci Instrum* 1966; **37**: 93–102.
8. Shaw TM, Elsken RH, Nuclear magnetic resonance absorption in hygroscopic materials. *J Chem Phys* 1950; **18**: 1113–1114.
9. Odebald E, Lindstrom G, Some preliminary observations on the proton magnetic resonance in biological samples. *Acta Radiol* 1955; **43**: 469–476.
10. Damadian R, Tumor detection by nuclear magnetic resonance. *Science* 1971; **171**: 1151–1153.
11. Moon RB, Richards JH, Determination of intracellular pH by ^{31}P magnetic resonance. *J Biol Chem* 1973; **248**: 7276–7278.
12. Hoult DI, Busby SJ, Gadian DG, Radda GK, Richards RE, Seeley PJ, Observation of tissue metabolites using ^{31}P nuclear magnetic resonance. *Nature* 1974; **252**: 285–287.
13. Chance B, Nakase Y, Bond M, Leigh JS Jr, McDonald G, Detection of ^{31}P nuclear magnetic resonance signals in brain by *in vivo* and freeze-trapped assays. *Proc Natl Acad Sci USA* 1978; **75**: 4925–4929.
14. Ackerman JJ, Grove TH, Wong GG, Gadian DG, Radda GK, Mapping of metabolites in whole animals by ^{31}P NMR using surface coils. *Nature* 1980; **283**: 167–170.
15. Behar KL, den Hollander JA, Stromski ME, Ogino T, Shulman RG, Petroff OA, Prichard JW, High-resolution ^{1}H nuclear magnetic resonance study of cerebral hypoxia *in vivo*. *Proc Natl Acad Sci USA* 1983; **80**: 4945–4948.
16. Alger JR, Sillerud LO, Behar KL, Gillies RJ, Shulman RG, Gordon RE, Shae D, Hanley PE, *In vivo* carbon-13 nuclear magnetic resonance studies of mammals. *Science* 1981; **214**: 660–662.
17. Lauterbur PC, Image formation by induced local interactions: examples employing nuclear magnetic resonance. *Nature* 1973; **242**: 190–191.
18. Mansfield P, Multiplanar image formation using NMR spin echoes. *J Phys C: Solid State Phys* 1977; **10**: L55–L58.
19. de Graaf RA, *In vivo NMR Spectroscopy. Principles and Techniques*. Chichester: Wiley, 1998.
20. Pan JW, Hamm JR, Rothman DL, Shulman RG, Intracellular pH in human skeletal muscle by ^{1}H NMR. *Proc Natl Acad Sci USA* 1988; **85**: 7836–7839.
21. Yoshizaki K, Seo Y, Nishikawa H, High-resolution proton magnetic resonance spectra of muscle. *Biochim Biophys Acta* 1981; **678**: 283–291.
22. Damon BM, Hsu AC, Stark HJ, Dawson MJ, The carnosine C-2 proton's chemical shift reports intracellular pH in oxidative and glycolytic muscle fibers. *Magn Reson Med* 2003; **49**: 233–240.
23. Schick F, Eismann B, Jung WI, Bongers H, Bunse M, Lutz O, Comparison of localized proton NMR signals of skeletal muscle and fat tissue *in vivo*: two lipid compartments in muscle tissue. *Magn Reson Med* 1993; **29**: 158–167.
24. Boesch C, Slotboom J, Hoppeler H, Kreis R, *In vivo* determination of intra-myocellular lipids in human muscle by means of localized ^{1}H-MR-spectroscopy. *Magn Reson Med* 1997; **37**: 484–493.
25. Szczepaniak LS, Babcock EE, Schick F, Dobbins RL, Garg A, Burns DK, McGarry JD, Stein DT, Measurement of intracellular triglyceride stores by H spectroscopy: validation *in vivo*. *Am J Physiol* 1999; **276**: E977–989.
26. Hwang JH, Pan JW, Heydari S, Hetherington HP, Stein DT, Regional differences in intramyocellular lipids in humans observed by *in vivo* ^{1}H-MR spectroscopic imaging. *J Appl Physiol* 2001; **90**: 1267–1274.
27. Steidle G, Machann J, Claussen CD, Schick F, Separation of intra- and extramyocellular lipid signals in proton MR spectra by determination of their magnetic field distribution. *J Magn Reson* 2002; **154**: 228–235.
28. Boesch C, Kreis R, Dipolar coupling and ordering effects observed in magnetic resonance spectra of skeletal muscle. *NMR Biomed* 2001; **14**: 140–148.
29. Krssak M, Petersen KF, Bergeron R, Price T, Laurent D, Rothman DL, Roden M, Shulman GI, Intramuscular glycogen and intramyocellular lipid utilization during prolonged exercise and recovery in man: a ^{13}C and ^{1}H nuclear magnetic resonance spectroscopy study. *J Clin Endocrinol Metab* 2000; **85**: 748–754.
30. Sinha R, Dufour S, Petersen KF, LeBon V, Enoksson S, Ma YZ, Savoye M, Rothman DL, Shulman GI, Caprio S, Assessment of skeletal muscle triglyceride content by (1)H nuclear magnetic resonance spectroscopy in lean and obese adolescents: relationships to insulin sensitivity, total body fat, and central adiposity. *Diabetes* 2002; **51**: 1022–1027.

31. Krssak M, Falk Petersen K, Dresner A, DiPietro L, Vogel SM, Rothman DL, Roden M, Shulman GI, Intramyocellular lipid concentrations are correlated with insulin sensitivity in humans: a ^{1}H NMR spectroscopy study. *Diabetologia* 1999; **42**: 113–116.
32. Wang ZY, Noyszewski EA, Leigh JS Jr, *In vivo* MRS measurement of deoxymyoglobin in human forearms. *Magn Reson Med* 1990; **14**: 562–567.
33. Tran TK, Sailasuta N, Hurd R, Jue T, Spatial distribution of deoxymyoglobin in human muscle: an index of local tissue oxygenation. *NMR Biomed* 1999; **12**: 26–30.
34. Kreis R, Bruegger K, Skjelsvik C, Zwicky S, Ith M, Jung B, Baumgartner I, Boesch C, Quantitative (1)H magnetic resonance spectroscopy of myoglobin de- and reoxygenation in skeletal muscle: reproducibility and effects of location and disease. *Magn Reson Med* 2001; **46**: 240–248.
35. Ugurbil K, Magnetization transfer measurements of individual rate constants in the presence of multiple reactions. *J Magn Reson* 1985; **64**: 207–219.
36. Price TB, Rothman DL, Avison MJ, Buonamico P, Shulman RG, ^{13}C-NMR measurements of muscle glycogen during low-intensity exercise. *J Appl Physiol* 1991; **70**: 1836–1844.
37. Shulman GI, Cline G, Schumann WC, Chandramouli V, Kumaran K, Landau BR, Quantitative comparison of pathways of hepatic glycogen repletion in fed and fasted humans. *Am J Physiol* 1990; **259**: E335–341.
38. Carey PE, Halliday J, Snaar JE, Morris PG, Taylor R, Direct assessment of muscle glycogen storage after mixed meals in normal and type 2 diabetic subjects. *Am J Physiol Endocrinol Metab* 2003; **284**: E688–694.
39. Shulman GI, Rothman DL, Jue T, Stein P, DeFronzo RA, Shulman RG, Quantitation of muscle glycogen synthesis in normal subjects and subjects with non-insulin-dependent diabetes by ^{13}C nuclear magnetic resonance spectroscopy. *New Engl J Med* 1990; **322**: 223–228.
40. van der Veen JW, de Beer R, Luyten PR, van Ormondt D, Accurate quantification of *in vivo* ^{31}P NMR signals using the variable projection method and prior knowledge. *Magn Reson Med* 1988; **6**: 92–98.
41. Maintz D, Heindel W, Kugel H, Jaeger R, Lackner KJ, Phosphorus-31 MR spectroscopy of normal adult human brain and brain tumours. *NMR Biomed* 2002; **15**: 18–27.
42. van den Boogaart A, Howe FA, Rodrigues LM, Stubbs M, Griffiths JR, *In vivo* ^{31}P MRS: absolute concentrations, signal-to-noise and prior knowledge. *NMR Biomed* 1995; **8**: 87–93.
43. de Graaf AA, Bovee WM, Improved quantification of *in vivo* ^{1}H NMR spectra by optimization of signal acquisition and processing and by incorporation of prior knowledge into the spectral fitting. *Magn Reson Med* 1990; **15**: 305–319.
44. Provencher SW, Estimation of metabolite concentrations from localized *in vivo* proton NMR spectra. *Magn Reson Med* 1993; **30**: 672–679.
45. Pfeuffer J, Tkac I, Provencher SW, Gruetter R, Toward an *in vivo* neurochemical profile: quantification of 18 metabolites in short-echo-time ^{1}H NMR spectra of the rat brain. *J Magn Reson* 1999; **141**: 104–120.
46. Sillerud LO, Shulman RG, Structure and metabolism of mammalian liver glycogen monitored by carbon-13 nuclear magnetic resonance. *Biochemistry* 1983; **22**: 1087–1094.
47. Zang LH, Laughlin MR, Rothman DL, Shulman RG, 13C NMR relaxation times of hepatic glycogen *in vitro* and *in vivo*. *Biochemistry* 1990; **29**: 6815–6820.
48. Chen W, Zhu XH, Avison MJ, Shulman RG, Nuclear magnetic resonance relaxation of glycogen H1 in solution. *Biochemistry* 1993; **32**: 9417–9422.
49. Murphy E, Hellerstein M, Is *in vivo* nuclear magnetic resonance spectroscopy currently a quantitative method for whole-body carbohydrate metabolism? *Nutr Rev* 2000; **58**: 304–314.
50. Overloop K, Vanstapel F, Van Hecke P, ^{13}C-NMR relaxation in glycogen. *Magn Reson Med* 1996; **36**: 45–51.
51. Overloop K, Van Hecke P, Vanstapel F, Chen H, Van Huffel S, Knijn A, van Ormondt D, Evaluation of signal processing methods for the quantification of a multi-exponential signal: the glycogen ^{13}C-1 NMR signal. *NMR Biomed* 1996; **9**: 315–321.
52. Roser W, Beckmann N, Wiesmann U, Seelig J, Absolute quantification of the hepatic glycogen content in a patient with glycogen storage disease by ^{13}C magnetic resonance spectroscopy. *Magn Reson Imag* 1996; **14**: 1217–1220.
53. Gruetter R, Prolla TA, Shulman RG, ^{13}C NMR visibility of rabbit muscle glycogen *in vivo*. *Magn Reson Med* 1991; **20**: 327–332.

54. Taylor R, Price TB, Rothman DL, Shulman RG, Shulman GI, Validation of ^{13}C NMR measurement of human skeletal muscle glycogen by direct biochemical assay of needle biopsy samples. *Magn Reson Med* 1992; **27**: 13–20.
55. Shalwitz RA, Reo NV, Becker NN, Ackerman JJ, Visibility of mammalian hepatic glycogen to the NMR experiment, *in vivo*. *Magn Reson Med* 1987; **5**: 462–465.
56. Rothman DL, Magnusson I, Katz LD, Shulman RG, Shulman GI, Quantitation of hepatic glycogenolysis and gluconeogenesis in fasting humans with ^{13}C NMR. *Science* 1991; **254**: 573–576.
57. Gruetter R, Magnusson I, Rothman DL, Avison MJ, Shulman RG, Shulman GI, Validation of ^{13}C NMR measurements of liver glycogen *in vivo*. *Magn Reson Med* 1994; **31**: 583–588.
58. Zang LH, Rothman DL, Shulman RG, ^{1}H NMR visibility of mammalian glycogen in solution. *Proc Natl Acad Sci USA* 1990; **87**: 1678–1680.
59. Kunnecke B, Seelig J, Glycogen metabolism as detected by *in vivo* and *in vitro* ^{13}C-NMR spectroscopy using [1,2-$^{13}C2$]glucose as substrate. *Biochim Biophys Acta* 1991; **1095**: 103–113.
60. Brainard JR, Hutson JY, Hoekenga DE, Lenhoff R, Ordered synthesis and mobilization of glycogen in the perfused heart. *Biochemistry* 1989; **28**: 9766–9772.
61. Zang LH, Howseman AM, Shulman RG, Assignment of the ^{1}H chemical shifts of glycogen. *Carbohydr Res* 1991; **220**: 1–9.
62. Chen W, Avison MJ, Zhu XH, Shulman RG, NMR studies of ^{1}H NOEs in glycogen. *Biochemistry* 1993; **32**: 11483–11487.
63. Chen W, Avison MJ, Bloch G, Shulman RG, Zhu XH, Proton NMR observation of glycogen *in vivo*. *Magn Reson Med* 1994; **31**: 576–579.

3

Metabolic Control Analysis for the NMR Spectroscopist

David A. Fell

School of Biological and Molecular Sciences, Oxford Brookes University, Gipsy Lane, Oxford OX3 0BP, UK

3.1. INTRODUCTION

The fundamental premise in studying intermediary metabolism, including metabolomics, is that activities of pathways are dependent upon the kinetics of the constituent enzymes. Hence when the pattern of metabolism changes in response to the needs of the cell, or the organism, this must be attributable to changes in the activities or kinetics of some of the enzymes. However, in order to make specific links between the

Metabolomics by In Vivo NMR. Edited by R. G. Shulman and D. L. Rothman
 ISBN: 0-470-84719-0

metabolic and enzymic changes, a theoretical framework is needed. For much of the twentieth century, explanations of the control of metabolic rate focussed on supposed rate-limiting steps. These were identified as irreversible enzymes near the start of a metabolic pathway, typically exhibiting cooperative kinetics and feedback inhibition by the products of the pathway, and sometimes undergoing covalent modification reactions that responded to external signals. This conceptual approach led to an experimental concentration on the kinetic and regulatory properties of selected enzymes, and on the measurement of changes in the concentration of key metabolites when the rate of metabolism was altered. Whilst this undoubtedly uncovered many details of the regulatory mechanisms in metabolism, there was no quantitative framework for deciding whether the effects observed were sufficient to account for the changes in metabolic rate. Consequently, there was no means of deciding between the hypotheses advanced by workers who favoured different enzymes as the rate-limiting step of a given metabolic process.

This changed with the development of metabolic control analysis by Kacser and Burns (Kacser and Burns, 1973; Kacser *et al.*, 1995) in Edinburgh and Heinrich and Rapoport (1974a, b) in Berlin. Instead of assuming *a priori* that control lay with a single key enzyme, this theory proposed that control of the metabolic rate (or flux as it is termed in this context) could be distributed over a number of steps, and the degree of control by individual enzymes could be quantitatively characterized by the values of appropriate coefficients. A specific prediction of the theory was that feedback-inhibited enzymes, far from being rate-limiting, would be poor sites of control. This conclusion was not readily accepted, in spite of measurements confirming it, but it was vindicated when attempts at changing metabolic fluxes using genetic manipulations to change the expression levels of such enzymes failed to have any effect (e.g. Schaaff *et al.*, 1989). Of course, this raises the question of what purpose is served by cooperativity and feedback inhibition, and this issue will be revisited elsewhere in this chapter.

The theory of metabolic control analysis developed rapidly through the 1980s and 1990s, but experimental techniques lagged behind that could measure the coefficients needed to make a quantitative assessment of control in a particular metabolic state. It is here that NMR has played a role, since one route to the answers is to measure both the change in metabolic flux and the changes in metabolite concentrations as an original metabolic state is altered to a new one. As will be shown later, NMR enables both these types of measurement to be made *in vivo*, thus allowing quantitation of control and regulation in a more natural physiological setting than is possible in many of the other available experimental approaches. This, therefore, is the justification for giving the following summary of the relevant aspects of the theory of metabolic control analysis. The treatment here is necessarily brief; I have given more extended coverage of many of the points in my review (Fell, 1992) and book (Fell, 1997) on the topic.

3.2. CONTROL COEFFICIENTS AND ELASTICITIES

3.2.1. The Flux Control Coefficient

The starting point for metabolic control analysis is to define measures of the effect that a change in the amount or activity of an enzyme will have on the steady-state value of a metabolic variable, such as the pathway flux. Both from the initial theory and many subsequent experiments, it is known that this is a continuously varying relationship: in many cases, the metabolic flux has a roughly hyperbolic dependence on the enzyme activity, as in the example in Figure 3.1. Metabolic control analysis therefore uses scaled sensitivity coefficients to measure the effects of an individual enzyme on the flux and other metabolic variables at a specific level of enzyme activity. When the variable is a flux, J, then the measure is called the 'flux control coefficient' (Burns *et al.*, 1985), where the coefficient C^J_{xase} for the effect of enzyme *xase*

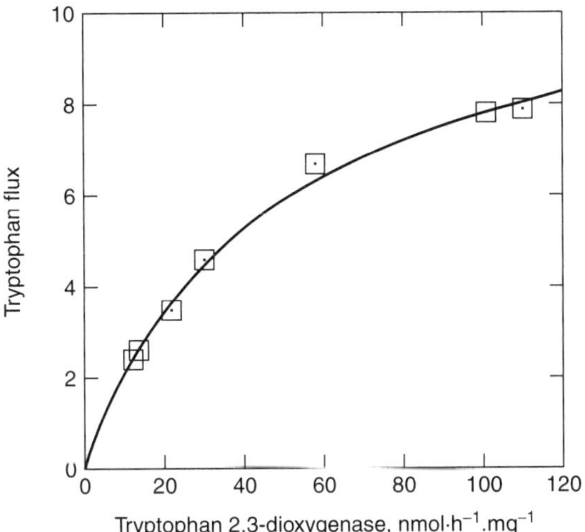

Figure 3.1. Dependence of tryptophan metabolism flux on tryptophan dioxygenase. The enzyme activity was experimentally varied in rats by Salter *et al.* (1986).

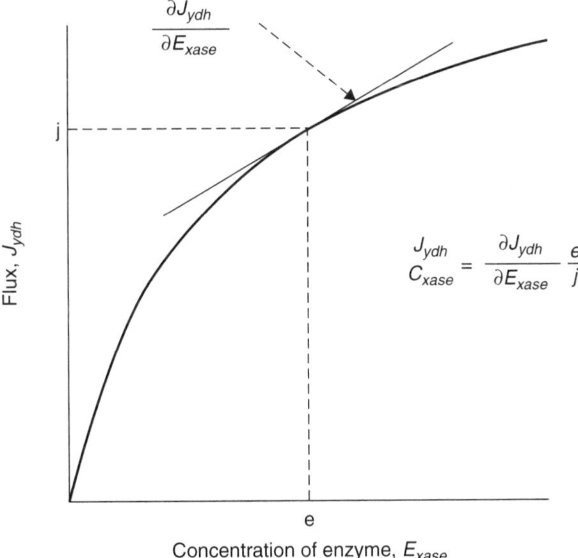

Figure 3.2. Definition of the flux control coefficient.

on the flux is the fractional (or percentage) change in flux caused by a fractional (or percentage) change in the enzyme, all other enzyme activities remaining constant:

$$C^{J}_{xase} = \frac{\partial J}{\partial E_{xase}} . \frac{E_{xase}}{J} = \frac{\partial \ln J}{\partial \ln E_{xase}} \tag{3.1}$$

This can be visualized as the scaled slope of the tangent of the flux–enzyme curve at the enzyme activity level being considered, as illustrated in Figure 3.2. The flux control coefficient approximately represents the percentage change in flux that would be caused by a 1 % change in the amount or activity of the enzyme. When more than one flux can be observed in the metabolism under study (because of metabolic branch points), then theory predicts that the flux control coefficients of the enzyme are likely to be different for each of the fluxes. Although the flux control coefficient is defined in terms of the effect observed when the amount or activity of the enzyme is altered, it has wider application than this, since it will still be a major determinant of the response of the flux to an effector that alters the activity of the enzyme by altering its K_m for its substrate [see equations (3.11) and (3.12)].

Kacser and Burns (1973) discovered a useful constraint on the values of the flux control coefficients: the flux summation theorem. This states that, if one takes the flux control coefficients of all n enzymes in the metabolic system, their sum will be 1:

$$\sum_{i=1...n} C^J_{E_i} = 1 \tag{3.2}$$

In a linear pathway with usual types of enzyme kinetics, which results in all flux control coefficients being 0 or positive, this theorem limits the value of a single flux control coefficient to a maximum of 1, provided that all the others are 0. In this case, the enzyme with the flux control coefficient of 1 would be a rate-limiting step, since the metabolic flux would change in direct proportion to the activity of that enzyme alone. However, there is no requirement in metabolic control analysis that control is distributed in this way, and experimentally it has been an occasional, but not frequent, finding. More often, the control has been distributed between a number of enzymes, none of which are truly rate-limiting. In branched metabolic pathways, it is true that this limitation is not absolute, since some enzymes (those that drain flux away from the flux being considered) will have negative flux control coefficients. In principle, this would allow the existence of large positive flux control coefficients provided they were balanced by large negative ones. Although there are limited circumstances where this could arise, it is not usual, so even in branched pathways the largest flux control coefficients rarely seem to exceed 1.

The summation theorem also shows that the flux control coefficient of an enzyme is a property of the whole metabolic system, not of the enzyme alone. Consider Figure 3.3, which shows the value of

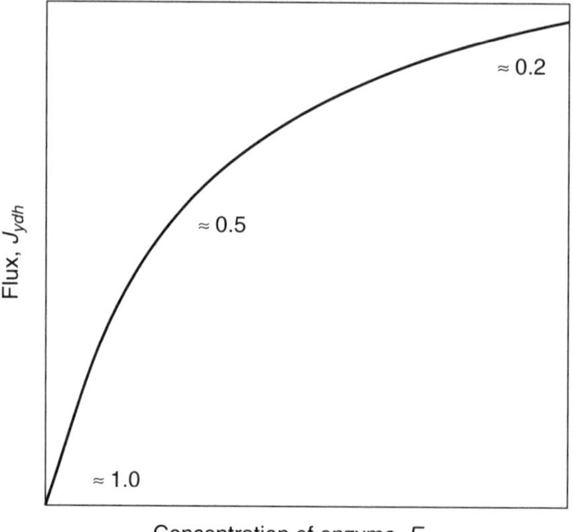

Figure 3.3. Values of the flux control coefficient.

an enzyme's flux control coefficient at various points along the flux–enzyme curve. Such curves can be, and have been, produced by placing the gene for the enzyme behind a variably inducible promoter or by titrating an enzyme with a specific inactivator (or noncompetitive inhibitor). Suppose that we start at the right-hand side of the graph at the normal level of activity of the enzyme in the cell, where the flux control coefficient is 0.2. At that point, the sum of the flux control coefficients of all the other enzymes is 0.8. As we move leftward on the x-axis, the flux control coefficient of the enzyme we are operating on rises to 0.5, but the sum of the flux control coefficients of all the others falls to 0.5. This continues as we move further to the left and the flux control coefficient of our enzyme approaches 1, whilst the sum of all the others approaches 0. Nothing has been done directly to any of the other enzymes, but their total flux control has varied between 0.8 and 0, showing that they alone do not determine the values of their own flux control coefficients: the value of a flux control coefficient is a property of the metabolic system as a whole.

3.2.2. Concentration Control Coefficients

When an enzyme activity changes in a metabolic pathway, it can have an impact on metabolite concentrations, tending to reduce the concentrations of metabolites upstream and increase those of the ones downstream. Except for those concentrations held constant either by the experimenter, the environment or physiological mechanisms outside the metabolic system, metabolite levels are variables of the system and their responses to a change in an enzyme activity can be characterized by concentration control coefficients, defined for any metabolite Y in the same way as the flux control coefficient:

$$C^{Y}_{xase} - \frac{\partial Y}{\partial E_{xase}} \cdot \frac{E_{xase}}{Y} = \frac{\partial \ln Y}{\partial \ln E_{xase}} \qquad (3.3)$$

Heinrich and Rapoport (1974a) discovered the summation theorem applying to concentration control coefficients: the sum of the coefficients of all the enzymes for any single metabolite is zero.

$$\sum_{i=1\ldots n} C^{Y}_{E_i} = 0 \qquad (3.4)$$

Since the positive coefficients are balanced by negative coefficients, the magnitude of any concentration control coefficient is less constrained, and they tend to be greater than flux control coefficients. The relative magnitude of an enzyme's flux and concentration control coefficients is an issue we will return to later.

3.2.3. Elasticity Coefficients

So far, the flux and concentration control coefficients have been treated as experimental observables, but a key achievement of the originators of metabolic control analysis (Kacser and Burns, 1973; Heinrich and Rapoport, 1974a) was to show that these control coefficients could be related to the kinetic properties of the enzymes in the pathway. The key to making this link is to express the kinetics in a form that is mathematically compatible with the control coefficients. This is achieved with the elasticity coefficient (Kacser and Burns, 1973; Burns *et al.*, 1985), which is a scaled sensitivity coefficient of the activity of an isolated enzyme to variation in the concentration of a metabolite. By *isolated enzyme*, we mean not that the enzyme needs to be extracted from the metabolic system, but that we are considering the response its activity would show if no metabolites changed in consequence of any change in its rate. Unlike usual measures of enzyme kinetics, the elasticity coefficient with respect to a specific metabolite is evaluated with all metabolites, including substrates, products and effectors, at the values they have in the vicinity of the enzyme in the metabolic state under consideration. Suppose an infinitesimal change, δY, is made in the amount of a metabolite Y that affects the rate of the reaction, v_{xase} catalysed by the enzyme *xase*, producing a change δv_{xase}. (The symbol v and term *rate* are used here rather than J and *flux* to emphasize it is the response of the isolated enzyme being considered, not of the pathway. Indeed, it is to mark this difference that metabolic control analysis employs this somewhat nonstandard definition of the term 'flux'.) The elasticity coefficient is then formally defined, in the limit as δY tends to zero, as:

$$\varepsilon_Y^{xase} = \frac{\partial v_{xase}}{\partial Y}.\frac{Y}{v_{xase}} = \frac{\partial \ln v_{xase}}{\partial \ln Y} \tag{3.5}$$

Values of the elasticity can be determined experimentally, although, if an adequate kinetic equation is available and the concentrations of all the metabolites are known, Equation (3.5) can be evaluated by differentiation of the rate equation. This is useful for relating elasticities to common enzyme kinetic functions (Groen *et al*., 1982). For example, for a Michaelis–Menten enzyme in the absence of product (even though this would not be a normal condition *in vivo*), the elasticity coefficient with respect to substrate goes from 1 at substrate concentrations well below the K_m value towards 0 as saturation is approached. Product inhibition elasticities are negative; for Michaelis–Menten kinetics far from equilibrium, values range from 0 to -1. Whereas traditional theory of metabolic regulation has tended to discount product inhibition unless it is strong enough to cause a significant diminution of the rate, metabolic control theory has shown that even very small product elasticities cause a system to behave very differently from one where product elasticities are zero (Fell and Sauro, 1985; Cornish-Bowden and Cárdenas, 2001); this has led to insufficient attention being given to the measurement of product effects on enzymes. For enzymes exhibiting cooperativity, the elasticities of substrates, products and effectors are less than or equal in magnitude to the instantaneous Hill coefficient. For enzymes that are close to equilibrium, the substrate and product elasticities approach plus and minus infinity, respectively. More details of the relationships between elasticities and conventional enzyme kinetics are given in Fell (1997).

The link between elasticities and control coefficients comes through the response coefficient. Suppose that a small change is made to the concentration of a metabolite, Y. If this metabolite only affects the activity of enzyme *xase*, then the pathway flux will be affected to an extent that depends on the flux control coefficient of this enzyme. The overall effect (Kacser and Burns, 1973) is that the relative *response* of the flux to the change in Y is given by the product of the flux control coefficient and the elasticity:

$$R_Y^J = C_{xase}^J \varepsilon_Y^{xase} \tag{3.6}$$

Where the metabolite Y is an effector that is external to the metabolic system under consideration (that is, it is neither synthesized nor consumed by the reactions of the system), the response coefficient indicates that there will only be an impact on the flux provided that there are significant, finite effects both of Y on *xase* (indicated by the elasticity) and of *xase* on the flux (indicated by the flux control coefficient). Although the former is a property of the enzyme alone, the latter is a property of the system itself. If Y is an effector that is specific to *xase*, or if *xase* is the first enzyme of the pathway and Y is its substrate, and it is possible to determine the value of its elasticity, then measuring the response of the metabolic flux to Y allows estimation of the flux control coefficient.

The result is different if Y is an internal metabolite of the system. In this case, there is usually more than one enzyme that is affected, in the sense of having a nonzero elasticity with respect to Y. The more important difference, though, is that Y is now a system variable and hence, even if it is altered from its steady-state value, the system will normally return to its initial state with unaltered fluxes, so the response is actually zero. Thus instead of Equation (3.6) we obtain the *flux connectivity theorem* (Kacser and Burns, 1973):

$$\sum_{i=1\ldots n} C_i^J \varepsilon_Y^i = 0 \tag{3.7}$$

Here we can take the sum over all the enzymes in the system because those that do not respond to Y will have a zero elasticity and hence will not contribute to the response. Similar relationships apply to the response of metabolite concentrations. When Y is an external metabolite, the response of the internal metabolite S is given by:

$$R_Y^S = C_{xase}^S \varepsilon_Y^{xase} \tag{3.8}$$

When Y is an internal metabolite, as with metabolic flux, there is no long-term response to a change in its concentration, and so we get, by analogy with Equation (3.7), the *concentration connectivity theorem* (Chen and Westerhoff, 1986), except that this takes two forms. The first is when the concentration control coefficients are those of a different metabolite to Y, when we have:

$$\sum_{i=1...n} C_i^S \varepsilon_Y^i = 0 \tag{3.9}$$

In the case of the concentration control coefficients of Y itself, however, the equation takes the form:

$$\sum_{i=1...n} C_i^Y \varepsilon_Y^i = -1 \tag{3.10}$$

This can be interpreted as meaning that, if the steady-state value of Y is perturbed, it will induce an equal and opposite response that returns Y to its original value.

3.2.4. Response to a Change in K_m

The response of a pathway to a change in the K_m (or $S_{0.5}$) of an enzyme is useful to consider since covalent modification of enzymes and effectors of allosteric enzymes act in this way. It is worth noting first of all that, in the case of an effector that changes the activity of an enzyme by altering the K_m, the response of the pathway is as given in the previous section. There is no difference in the response coefficient whether the target enzyme has its catalytic activity changed directly by the effector or whether its substrate affinity is changed; both effects are reflected by the elasticity coefficient, which is based on the change in activity at fixed concentrations of all metabolites except one. (There may well be a difference between these two cases in the response to large changes in the effector, but they will not be apparent for small changes.)

Where the kinetic characteristics are changed by, for example, phosphorylation of an enzyme, then we do need to know the response to a change in K_m. It is possible to define an elasticity of an enzyme with respect to one of its K_m values; the definition is no different from the standard definition. We would therefore expect the flux response to a change in the K_m of enzyme *xase* to be:

$$R_{K_m}^J = C_{xase}^J \varepsilon_{K_m}^{xase} \tag{3.11}$$

However, since most enzyme kinetic equations can be written in such a way that every K_m appears as the denominator in a term that contains the appropriate substrate S as the numerator, it follows from the definition of the elasticity that

$$\varepsilon_{K_m}^{xase} = -\varepsilon_S^{xase}$$

Hence

$$R_{K_m}^J = -C_{xase}^J \varepsilon_S^{xase} \tag{3.12}$$

This applies even if S itself is an internal metabolite of the system, since the K_m is a parameter in this context.

3.2.5. Relating Control Coefficients and Elasticities

The summation and connectivity theorems between them allow the control coefficients to be expressed in terms of the elasticities of the enzymes. This can be illustrated for a simple two-step, linear pathway, although the approach can be generalized to more realistic cases (Fell and Sauro, 1985; Reder, 1988):

$$X_0 \xrightarrow{xase} Y \xrightarrow{ydh} X_1 \qquad \text{(scheme 3.1)}$$

This has the flux summation theorem and flux connectivity equations:

$$\begin{aligned} C^J_{xase} + C^J_{ydh} &= 1 \\ C^J_{xase}\varepsilon^{xase}_Y + C^J_{ydh}\varepsilon^{ydh}_Y &= 0 \end{aligned} \qquad (3.13)$$

Expressing the control coefficients in terms of the elasticities gives:

$$\begin{aligned} C^J_{xase} &= \frac{\varepsilon^{ydh}_Y}{\varepsilon^{ydh}_Y - \varepsilon^{xase}_Y} \\ C^J_{ydh} &= \frac{-\varepsilon^{xase}_Y}{\varepsilon^{ydh}_Y - \varepsilon^{xase}_Y} \end{aligned} \qquad (3.14)$$

Simple though the pathway is, the result is representative in that it shows that the expressions for the control coefficients contain elasticities for all the enzymes in the system. Fell and Sauro (1985) developed a method of extending this type of analysis to more complex pathways, and they and other authors subsequently generalized the computation of these relationships still further. One of the first experimental determinations of control coefficients from elasticities was by Groen *et al.* (1986) for 11 steps of hepatocyte gluconeogenesis, and this remains one of the largest pathways analysed in this way.

The corresponding result for the concentration control coefficients follows from the concentration summation theorem and the corresponding connectivity theorem:

$$\begin{aligned} C^Y_{xase} + C^Y_{ydh} &= 0 \\ C^Y_{xase}\varepsilon^{xase}_Y + C^Y_{ydh}\varepsilon^{ydh}_Y &= -1 \end{aligned} \qquad (3.15)$$

This leads to:

$$\begin{aligned} C^Y_{xase} &= \frac{1}{\varepsilon^{ydh}_Y - \varepsilon^{xase}_Y} \\ C^Y_{ydh} &= \frac{-1}{\varepsilon^{ydh}_Y - \varepsilon^{xase}_Y} \end{aligned} \qquad (3.16)$$

Comparing the results of Equations (3.14) and (3.16) we can see that both the flux and concentration control coefficients all have the same denominator term and differ only in the numerators. Hence not only can we know the relative values of the flux control coefficients of the different enzymes in the pathway, but, if we take a single enzyme, we can also know the relative values of its flux and concentration control coefficients. This shows that, if we make a change in the activity of enzyme *xase*, the impacts on flux and metabolite concentrations are coupled and have precise relative relationships. Therefore any explanation of how a metabolic pathway changes its flux must also account for any changes in metabolite concentrations. Later in the chapter it will become clear that changing metabolic rate, whilst more or less preserving metabolite homeostasis, as cells commonly do, is not a simple issue and needs specific explanation.

In the meantime, these results can now be used as the basis for illustrating further aspects of control analysis.

3.3. TOP-DOWN CONTROL ANALYSIS

Evidently there are not many two-step metabolic pathways of interest, but Kacser and Burns (1979) pointed out that a longer pathway could in principle be reduced to this form by regarding what were termed *xase* and *ydh* as groups of enzymes. Fell and Sauro (1985) showed how to relate the elasticities of groups of enzymes to those of the individual enzymes in the group; the control coefficients of the group are simply the sum of the control coefficients of the individual components (with the interpretation that the activities of all enzymes in the group are assumed to change in parallel). The main restriction to simplifying pathways in this way is that any metabolite that is contained within the group should not have interactions with any enzymes outside the group. (This restriction can be relaxed, but at the expense of making the group elasticities to the remaining common metabolites of the simplified pathway more difficult to interpret; Ainscow and Brand 1999).

The experimental determination of control coefficients from elasticities was simplified by the development of the top-down protocol by Brown *et al.* (1990). The concept was that the experimental approach is easier if taken incrementally, dividing the metabolic system first into a few large multienzyme blocks, rather than starting from the elasticities of every enzyme in a bottom-up manner. After an initial analysis which identifies how the control is distributed, it is always possible to divide the system into smaller blocks, or a similar number of blocks but at different points, in order to reach a finer level of detail. The issue is how to determine the group elasticities, and this is simple in the case of two blocks (or of a simple branched system of three blocks around a single common metabolite). Consider that, in the previous two-step Scheme 3.1, *xase* and *ydh* stood for two enzyme groups. If we had some way of specifically altering the activity of *xase*, then we would expect this to cause measurable changes in the flux through the pathway and in the concentration of Y. In their initial experiments on mitochondrial oxidative phosphorylation, Brown *et al.* (1990) exploited the highly specific inhibitors that exist for this pathway, but in Scheme 3.1 it would also be possible to change the activity of *xase* by altering the concentration of the input metabolite X_0. It is not necessary to know the amount by which the activity of *xase* is changed, because the aim of the experiment is to determine the concentration of Y at each flux value. Since the flux J is equal to the rate of *ydh*, a plot of $J-Y$ or, better, $\ln J - \ln Y$ gives ε_Y^{ydh}, the elasticity of *ydh* to Y. The procedure is illustrated in Figure 3.4.

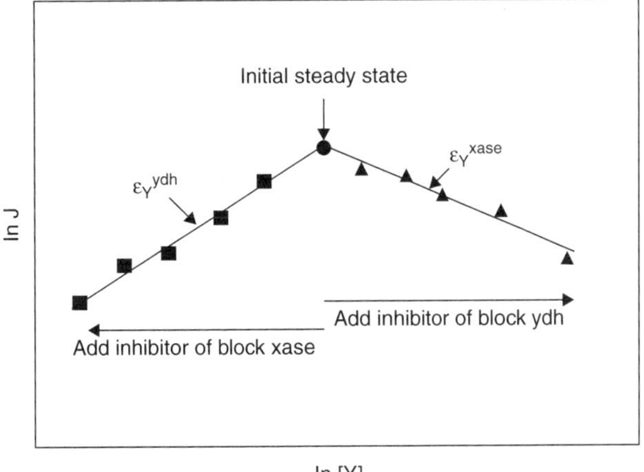

Figure 3.4. Top-down titration.

The next step in the experiment is to repeat it with the blocks exchanged, i.e. a selective inhibitor of *ydh* is used to alter the activity of this block, and measurements of the pathway and flux and the concentrations of Y allow estimation of the elasticity of block *xase*. The flux control coefficients can then be calculated from the two elasticities with Equation (3.14), and the concentration control coefficients from Equation (3.16).

3.4. SUPPLY AND DEMAND

The two block pathway of Scheme 3.1 can be regarded as the archetype of a supply–demand system for Y. For example, Y might represent a monomer used for making cell biomass, such as an amino acid or a nucleotide, in which *xase* represents the biosynthetic pathway for Y, and *ydh* represents the processes, such as protein or nucleic acid synthesis, that depend on Y. Another case would be where *xase* represents the conversion of nutrient X_0 to a common cellular intermediate Y, and *ydh* represents the synthesis of a storage compound such as glycogen. Traditional approaches to metabolic regulation have tended to focus on the supply and demand processes as separate issues, which has resulted in partial and rather misleading analyses. A focus on the supply–demand system as a whole, as has been carried out by Hofmeyr and Cornish-Bowden (1991, 2000) is more revealing, and shows how the two examples quoted above might be expected to have different designs. In the case of biosynthesis of a monomer, it seems reasonable to assume that it is advantageous for the flux to be controlled by the demand, and to exhibit little sensitivity to fluctuations in the concentration of the nutrient X_0. On the other hand, in the second case, there is no fixed level of demand for the storage compound, and its synthesis should respond to the availability of the input nutrient. Translated into control analysis terms, this implies that the response coefficient, $R^J_{X_0}$, of the flux to X_0 should be smaller in the former case and larger in the latter, which, given Equation (3.6), implies that control by the supply would be disadvantageous in the first case but advantageous in the second. In both cases, internal homeostasis of the cell is best served by the concentration of Y remaining relatively constant, i.e. small values for the concentration control coefficients of Y with respect to supply and demand (C^Y_{xase} and C^Y_{ydh}).

From the connectivity theorem for the two-block pathway, Equation (3.13), the relative sizes of the two control coefficients are given by

$$\frac{C^J_{xase}}{C^J_{ydh}} = \frac{\varepsilon^{ydh}_Y}{-\varepsilon^{xase}_Y}$$

(Note that the elasticity of *xase* to Y is a product elasticity and therefore itself negative. Furthermore, although it has not been shown here, if block *xase* contains an enzyme at the beginning that is feedback-inhibited by Y, the feedback elasticity may be the major component of ε^{xase}_Y.) Obviously, $C^J_{xase} > C^J_{ydh}$ when $\varepsilon^{ydh}_Y > |\varepsilon^{xase}_Y|$, which is obtained if product and feedback inhibition of Y on *xase* is weak, but Y has a strong activating effect on *ydh*, as obtained with positive substrate cooperativity. In this case, activation of the supply block can be used to increase the flux, but what is the effect on the homeostasis of Y? If we pose the question in the form: 'does a change in activity of *xase* have a larger effect on the flux J than on Y?', then it is useful to have a direct comparison of the size of the two control coefficients. This is given by the co-response coefficient (Hofmeyr *et al.*, 1993; Cornish-Bowden and Hofmeyr, 1994), defined as:

$$O^{J,Y}_{xase} = \frac{C^J_{xase}}{C^Y_{xase}} = \frac{\partial \ln J}{\partial \ln Y} \tag{3.17}$$

In this case, from the results in Equations (3.14) and (3.16), it follows that the co-response coefficient is the ratio of the numerators of the two control coefficients, since it has already been mentioned that they have identical denominators, giving:

$$O^{J,Y}_{xase} = \varepsilon^{ydh}_{Y} \tag{3.18}$$

Hence when changing the activity of *xase*, a large value for the elasticity of *ydh* with respect to Y ensures that the change in flux will be larger than the change in Y, although not by more than a typical Hill coefficient for substrate cooperativity. [If it seems strange that this co-response to *xase* should be entirely represented by a property of *ydh*, note that Equation (3.18) actually expresses the top-down procedure for determining the elasticity of *ydh* by an inhibitor titration of *xase* described in the previous section.]

The second case, control of the flux by the demand block, occurs when $\varepsilon^{ydh}_{Y} < |\varepsilon^{xase}_{Y}|$. If the demand block is activated or inhibited to change the flux, the relative effects on the flux and on Y are given by the co-response coefficient to *ydh*, which, from Equations (3.14) and (3.16) as before, gives:

$$O^{J,Y}_{ydh} = \varepsilon^{xase}_{Y} \tag{3.19}$$

This shows that the greater the sensitivity of the supply block to its product, the larger the change in flux is relative to the change in Y. Thus, although control can be transferred to the demand block by a very small value of ε^{ydh}_{Y}, such as would be obtained if the demand block was near-saturation with Y, that solution does not necessarily result in good homeostasis of Y when the flux changes. Homeostasis of Y requires high sensitivity of the supply block to Y, which can be obtained if Y is a strongly cooperative feedback inhibitor of the block. Again, the Hill coefficient of the inhibition is a likely upper limit of the size of the elasticity. (Note that the elasticity itself is negative, since an increase in flux by activating the demand is likely to lower Y.) Hofmeyr and Cornish-Bowden (1991) have argued convincingly that cooperativity in feedback inhibition has evolved to ensure metabolite homeostasis, since noncooperative inhibition would perform perfectly adequately to ensure that a metabolic system attained steady state and that the flux was controllable.

3.5. MULTISITE MODULATION AND PROPORTIONAL ACTIVATION

The relatively low values generally found for the flux control coefficients of single enzymes pose a significant problem: how do cells achieve large changes in flux? There is no doubt that the flux in some pathways can be changed by large factors over time scales too short to involve synthesis or degradation of the enzymes of the pathway; the increase in glycolysis from glycogen in muscle on contraction is just one example. Action on the activity of a single enzyme would not seem to be adequate, especially for activation of a pathway. The reason can be seen by returning to Figure 3.1; even if the enzyme activity starts at the left end of the curve, with the flux control coefficient in the range 1–0.5, any significant degree of activation moves the system to the right along the curve, where the flux control coefficient starts to fall, and the increase in flux for each further increment of enzyme activity becomes less and less. Metabolic control analysis cannot make an exact prediction of the increase in flux that will be obtained with a large increase in enzyme activity because the exact shape of the enzyme–flux curve is not defined. However, Small and Kacser (1993) derived an approximate formula that quantitatively supports the qualitative conclusion described here.

There is a further problem in understanding large flux changes in cells: it is apparent from the equations presented in the previous section that activation of a single enzyme will tend to cause changes in metabolite concentrations of the same order as the flux changes. Indeed, there are instances of just such an effect when single enzymes are overexpressed; for example, overexpression of phosphofructokinase in potato tubers causes no significant change in flux, but does increase the concentration of glycolytic metabolites downstream (Thomas *et al*., 1997). In physiological systems, such behaviour is far from usually the case;

some examples are reviewed in Fell (1997) and Fell and Thomas (1998), and they show that, in many metabolic systems, the flux changes are significantly larger than the changes in metabolite concentrations.

If control by a single enzyme is relatively limited in terms of the size of the flux response it can produce, and disadvantageous in terms of the adverse effects for metabolite homeostasis, what is the alternative? The answer was inherent in the original derivation of the flux summation theorem by Kacser and Burns (1973): they proposed that, if every enzyme in a metabolic system had its activity increase by exactly the same fractional amount, then the flux would increase in proportion, but the metabolite concentrations would remain undisturbed. This leads to the conclusion that the flux control coefficients total to 1 and the concentration control coefficients total to 0. In this case, all the fluxes in the system increase by the same amount. To achieve a selective increase in a particular target flux, whilst maintaining all metabolite levels constant, a more subtle combination of enzyme activity changes would be needed, as described by Kacser and Acerenza (1993) in their 'universal method' for flux increases. However, in this case as well, the enzyme activities in any linear segment of the metabolic network would all need to be increased in proportion. Since the metabolite concentrations do not change, there will be no changes in the elasticities of the enzymes, and hence no change in the control distribution. Therefore, the size of the flux change that can be obtained is not limited, as it is in the case of activation of a single enzyme.

In the case of long-term adjustments of metabolic flux, involving synthesis and/or degradation of enzymes, there is evidence that all the enzymes in a pathway adjust by similar (though not always identical) amounts. In bacteria, common segments of a pathway whose expression level varies in response to environmental signals are often found in an operon so that they are always co-expressed. In eukaryotes, where there are regulons rather than operons, nearly parallel changes in activity of all the enzymes in a pathway have often been found; I have previously reported a number of examples (Fell, 2000).

What about shorter time scales in metabolic control, before there is significant synthesis or degradation of pathway enzymes? It is clear that we know of no mechanisms that act in parallel on every enzyme of a metabolic pathway. However, it was pointed out that there were many known instances of pathways where there are control mechanisms, such as covalent modification of enzymes or the action of effectors, that act at several points along the pathway (Fell and Thomas, 1995; Fell, 1997; Thomas and Fell, 1998). Furthermore, the relative size of the flux changes relative to those in the metabolite concentrations are larger than would be expected from activation of a single controlling enzyme (Thomas and Fell, 1998), and therefore it must be concluded that the observed changes in the pathway must be the result of simultaneous operation of these controls at several points. We termed this *multisite modulation*. Korzeniewski *et al.* (1995) have produced both experimental and theoretical evidence for a similar control mechanism in energy metabolism. Firstly, they measured the changes in flux and the change in proton-motive force in mitochondria stimulated by vasopressin. By treating mitochondrial oxidative phosphorylation to a top-down analysis with the protonmotive force as the central metabolite Y, and measuring the co-response of respiratory flux and proton-motive force to the activation, they were able to calculate the relative degree of activation of the two blocks, on a scale where a *proportional activation* of one would signify equal degrees of activation of the two blocks. They concluded that both blocks were in fact equally activated by vasopressin. On the basis of computer simulations, Korzeniewski (2003) has also proposed that proportional activation must operate in muscle *in vivo* on both ATP utilization and oxidative phosphorylation during increases in aerobic energy production; this emerges as the most likely explanation of the size of the flux changes relative to changes in levels of metabolites such as ADP. The theory of measuring the degree of proportional activation and its application to glycogen metabolism in muscle is presented later in this book in Chapter 5.

Multisite modulation and proportional activation at the biochemical level could well be just an instance of a more general phenomenon. The concept of *symmorphosis* in mammalian physiological systems (Weibel *et al.* 1991; Weibel, 2000) is that limitations to performance are distributed throughout the system, and that upgrading the performance requires adjustments throughout in the structure and activity of the components.

3.6. CONCLUSION

The concept of the rate-limiting step as a general principle of metabolic control long held sway in biochemistry and is only now starting to disappear from textbooks. It was undermined on the theoretical front by metabolic control analysis, and on the experimental front partly by metabolic control analysis and partly by failed attempts at genetic engineering of metabolism. Metabolic control analysis has promoted two new concepts: that control can be shared between a number of steps, and that the total control that can be exerted is limited. Although these two concepts are gaining acceptance, their implications have not yet been fully absorbed. In essence, metabolic control analysis has been more successful in exposing the mechanisms that cannot work when cells are making significant changes in flux than in explaining how it actually happens. Since the control of flux and of concentrations can be distributed throughout a pathway, the answer must be that the mechanisms that change flux must act throughout the pathway. Thus all the molecular mechanisms of control are likely to be contributing, either to the flux change or to metabolite homeostasis, or both.

Under the influence of the rate-limiting step concept, it used to be thought necessary to identify one site of control as the critical point, and indeed it is still common for papers on enzymes (or their genes) to make claims for their rate-controlling role (usually gratuitously and without experimental support). Not only has this divisive and confrontational approach missed the point about the need for coordinated control at a number of sites, it has focussed too much attention on the aspect of flux control, and obscured the fact that some of the control machinery, such as feedback inhibition, is playing an important role in regulation of metabolite concentrations. Having said that, some of the cases examined in later in this book do show that there are specific instances where most of the flux control is concentrated in a single step. The difference here is that the effective rate-limiting step has been identified objectively by a procedure that did not have to show there was one at all. Furthermore, even where there is a rate-limiting step, there still needs to be an explanation of why control mechanisms act at other points in the pathway. Again this is likely to be for metabolite homeostasis, and also to ensure an unchanged flux control distribution as the activity of the flux-controlling enzyme changes.

Given that issues still remain in explaining how cells achieve control and regulation in large metabolic flux changes, there is great value in making simultaneous measurements of fluxes and metabolites *in vivo*, and it is here that NMR has much to offer.

REFERENCES

Ainscow EK, Brand MD (1999) Top-down control analysis of ATP turnover, glycolysis and oxidative phosphorylation in rat hepatocytes. *Eur J Biochem* **263**(3): 671.

Brown GC, Hafner RP, Brand MD (1990) A 'top-down' approach to the determination of control coefficients in metabolic control theory. *Eur J Biochem* **188**: 321–325.

Burns JA, Cornish-Bowden A, Groen AK, Heinrich R, Kacser H, Porteous JW, Rapoport SM, Rapoport TA, Stucki JW, Tager JM, Wanders RJA, Westerhoff HV (1985) Control analysis of metabolic systems. *Trends Biochem Sci* **10**: 16.

Chen YD, Westerhoff HV (1986) How do inhibitors and modifiers of individual enzymes affect steady-state fluxes and concentrations in metabolic systems? *Math Model* **7**: 1173–1180.

Cornish-Bowden A, Cárdenas ML (2001) Information transfer in metabolic pathways. Effects of irreversible steps in computer models. *Eur J Biochem* **268**: 6616–6624.

Cornish-Bowden A, Hofmeyr JHS (1994) Determination of control coefficients in intact metabolic systems. *Biochem J* **298**(Mar): 367–375.

Fell DA (1992) Metabolic control analysis: a survey of its theoretical and experimental developments. *Biochem J* **286**: 313–330.

Fell DA (1997) *Understanding the Control of Metabolism*. London: Portland Press.
Fell DA (2000) Signal transduction and the control of expression of enzyme activity. *Adv Enzym Regul* **40**: 35–46.
Fell DA, Sauro HM (1985) Metabolic control analysis: additional relationships between elasticities and control coefficients. *Eur J Biochem* **148**: 555–561.
Fell DA, Thomas S (1995) Physiological control of flux: the requirement for multisite modulation. *Biochem J* **311**: 35–39.
Fell DA, Thomas S (1998) Changes in enzyme expression and the control of metabolic flux. In *BioThermoKinetics in the Post-Genomics Era*, eds Larsson C, Pahlman IL, Gustafsson FE, Wijker JE, Kholodenko BN. Göteborg: Chalmers University of Technology, pp. 41–45.
Groen AK, van der Meer R, Westerhoff HV, Wanders RJA, Akerboom TPM, Tager JM (1982) Control of metabolic fluxes. In *Metabolic Compartmentation*, ed. Sies H. London: Academic Press, pp. 9–37.
Groen AK, van Roermund CWT, Vervoorn RC, Tager JM (1986) Control of gluconeogenesis in rat liver cells. *Biochem J* **237**: 379–389.
Heinrich R, Rapoport TA (1974a) A linear steady-state treatment of enzymatic chains; general properties, control and effector strength. *Eur J Biochem* **42**: 89–95.
Heinrich R, Rapoport TA (1974b) A linear steady state treatment of enzymatic chains. *Eur J Biochem* **42**: 97–105.
Hofmeyr JHS, Cornish-Bowden A (1991) Quantitative assessment of regulation in metabolic systems. *Eur J Biochem* **200**: 223–236.
Hofmeyr JHS, Cornish-Bowden A (2000) Regulating the cellular economy of supply and demand. *FEBS Lett* **476**: 47–51.
Hofmeyr JHS, Cornish-Bowden A, Rohwer JM (1993) Taking enzyme kinetics out of control: Putting control into regulation. *Eur J Biochem* **212**: 833–837.
Kacser H, Acerenza L (1993) A universal method for achieving increases in metabolite production. *Eur J Biochem* **216**: 361–367.
Kacser H, Burns JA (1973) The control of flux. *Symp Soc Exp Biol* **27**: 65–104. [Reprinted in *Biochem Soc Trans* **23**: 341–366 (1995).]
Kacser H, Burns JA (1979) Molecular democracy: who shares the controls? *Biochem Soc Trans* **7**: 1149–1160.
Kacser H, Burns JA, Fell DA (1995) The control of flux. *Biochem Soc Trans* **23**: 341–366.
Korzeniewski B (2003) Regulation of oxidative phosphorylation in different muscles and various experimental conditions. *Biochem J* **375**: 799–804.
Korzeniewski B, Harper ME, Brand MD (1995) Proportional activation coefficients during stimulation of oxidative phosphorylation by lactate and pyruvate or vasopressin. *Biochim Biophys Acta* **1229**: 315–322.
Reder C (1988) Metabolic control theory: a structural approach. *J Theor Biol* **135**: 175–201.
Salter M, Knowles RG, Pogson CI (1986) Quantification of the importance of individual steps in the control of aromatic amino acid metabolism. *Biochem J* **234**: 635–647.
Schaaff I, Heinisch J, Zimmerman FK (1989) Overproduction of glycolytic enzymes in yeast. *Yeast* **5**: 285–290.
Small JR, Kacser H (1993) Responses of metabolic systems to large changes in enzyme activities and effectors. 1. the linear treatment of unbranched chains. *Eur J Biochem* **213**: 613–624.
Thomas S, Fell DA (1998) Multiple enzyme activation in metabolic flux control. *Adv Enzyme Regul* **38**: 65–85.
Thomas S, Mooney PJF, Burrell MM, Fell DA (1997) Finite change analysis of lines of transgenic potato (*Solanum tuberosum*) overexpressing phosphofructokinase. *Biochem J* **322**: 111–117.
Weibel ER (2000) *Symmorphosis. On Form and Function in Shaping Life*. Cambridge, MA: Harvard University Press.
Weibel ER, Taylor CR, Hoppeler H (1991) The concept of symmorphosis: a testable hypothesis of structure-function relationship. *Proc Natl Acad Sci USA* **88**: 10357–10361.

4

MRS Studies of the Role of the Muscle Glycogen Synthesis Pathway in the Pathophysiology of Type 2 Diabetes

Gerald I. Shulman and Douglas L. Rothman

Howard Hughes Institute, Departments of Diagnostic Radiology and Biomedical Engineering, Yale University School of Medicine, New Haven, CT 06520-8056, USA

Metabolomics by In Vivo NMR. Edited by R. G. Shulman and D. L. Rothman
 ISBN: 0-470-84719-0

4.1. INTRODUCTION

The pathophysiology of type 2 diabetes (also known as adult-onset or non-insulin-dependent diabetes) involves defects in muscle and liver sensitivity to insulin and decreased insulin secretion. Normally under post-meal conditions the increase in plasma glucose sensed by the pancreas leads to enhanced insulin release. Insulin binds to insulin receptors on target organ cells, most prominently muscle, liver and heart, resulting in a series of cellular events that promote intracellular glucose transport and metabolism (1). Insulin resistance is defined as the inability of peripheral target tissues to respond properly to normal circulating concentrations of insulin. In insulin-resistant individuals at first the pancreas compensates by secreting increased amounts of insulin, allowing normal post-meal glucose concentrations to be maintained. Insulin resistance usually precedes the onset of type 2 diabetes by several decades, although recently a significant number of individuals have started developing it in their teens. Two factors contribute to the development of type 2 diabetes from the insulin resistant state. One factor is primarily environmental – plasma fatty acid levels increase due to inactivity and diet, resulting in enhanced insulin resistance. The elevated plasma fatty acid levels are often, but not always, associated with obesity. The increased insulin resistance leads to impaired glucose tolerance despite elevated insulin concentrations. Overt clinical type 2 diabetes results when an other factor, pancreatic beta cell failure, results in reduced insulin secretion. Often the increase in insulin resistance occurs in parallel with the reduction in pancreatic insulin secretion capacity.

This chapter focuses on studies performed over the last 16 years using MRS to study the role of muscle glycogen synthesis in insulin resistance associated with diabetes and obesity, and to determine the molecular mechanisms leading to this resistance. Because the MRS methodology and validation studies, as well as the control of muscle glycogen synthesis in the healthy state, are covered in earlier chapters, we will focus on studies of insulin resistance. In Section 4.2 the initial studies that established the importance of muscle glycogen synthesis in insulin resistance will be described. Section 4.3 covers the studies that found that the reduced muscle glycogen synthesis in individuals with a high genetic risk for insulin resistance leading to type 2 diabetes is in muscle glucose transport. Section 4.4 describes studies which determined the impact of environmental factors such as exercise and diet, particularly plasma lipid levels, on muscle glycogen synthesis and the pathway steps these factors act upon.

4.2. MRS STUDIES OF THE ROLE OF INSULIN-STIMULATED MUSCLE GLYCOGEN SYNTHESIS IN THE REDUCED GLUCOSE DISPOSAL CHARACTERISTIC OF INSULIN-RESISTANT STATES

Prior to the initial MRS studies of human glycogen synthesis, the role of insulin-stimulated muscle glycogen synthesis in the etiology of type 2 diabetes was controversial. Whole body balance studies had suggested that the major defect in glucose disposal in type 2 diabetes was glucose storage, with arterio-venous difference studies suggesting that the muscle was a major contributor (1). However conventional biopsy measurements had failed to establish a major role for glycogen, largely (as subsequent MRS studies showed) due to the loss of the small amount of glycogen synthesized during the studies during the time between removal and freezing of muscle biopsy samples. As described below, MRS studies were able to overcome this obstacle because MRS is a non-invasive technology and ^{13}C enrichment allows enhancement of the sensitivity of detection of very small percentage changes in glycogen levels. These studies have firmly established the important role of impaired muscle glycogen synthesis in insulin resistance.

4.2.1. Impaired Muscle Glycogen Synthesis During Hyperinsulinemic Hyperglycemia in Insulin-resistant Subjects and Healthy Subjects with Genetic Insulin Resistance

The initial studies using ^{13}C MRS to study diabetes were performed on subjects with type 2 diabetes and age-matched controls under conditions of hyperglycemic hyperinsulinemia, designed to mimic the plasma concentrations of glucose and insulin experienced by people with type 2 diabetes under post-prandial conditions (2). Hyperglycemic–hyperinsulinemic clamp studies were performed by infusing ^{13}C-enriched glucose and insulin intravenously and measuring the rate of muscle glycogen synthesis using surface coil MRS measurements of the calf (primarily the gastrocnemius and soleus muscles). The rate of muscle glycogen synthesis at 15 min intervals was determined using metabolic modeling. The rate was converted to whole body muscle glycogen synthesis based upon published measurements of whole body muscle mass (which subsequently were validated using MRI). The MRS determined rate was compared with the rate of whole-body glucose uptake and non-oxidative disposal of glucose, the latter determined by subtracting off the oxidative component of glucose consumption measured by indirect calorimetry. This study found that under these conditions muscle glycogen synthesis represented the predominant pathway for non-oxidative glucose disposal in normal subjects and that muscle glycogen synthesis was the major pathway of overall glucose metabolism. In addition, the rate of glycogen formation in subjects with diabetes was 60 % lower than the rate in normal subjects. The reduction in glycogen synthesis accounted quantitatively for their reduced whole body glucose uptake compared with controls. Based on data from this study, it was clear that impaired glycogen synthesis was the major intracellular metabolic defect responsible for insulin resistance in subjects with type 2 diabetes. Figure 4.1 compares the ^{13}C MRS time course of muscle glycogen synthesis in the subjects with type 2 diabetes and the control subjects.

Since this study was performed, a similar defect in muscle glycogen synthesis under hyperinsulinemic–hyperglycemic conditions has been measured in a variety of important insulin-resistant states including individuals with a high genetic risk of type 2 diabetes, obese subjects and subjects with type 1 diabetes. These studies will be discussed in the context of the molecular defects leading to insulin resistance in later sections of this chapter.

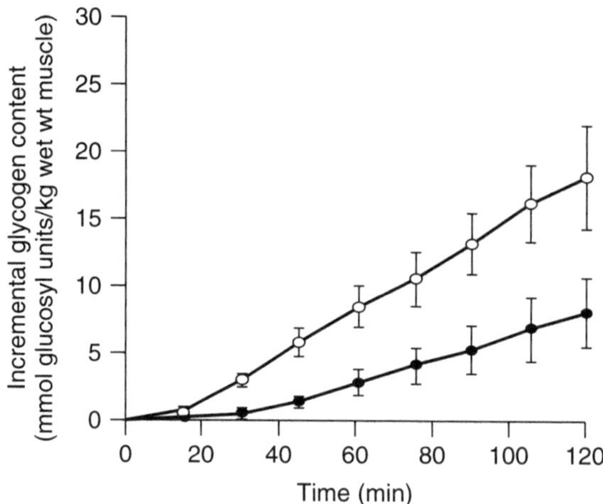

Figure 4.1. Time course of muscle glycogen concentration calculated from the ^{13}C MRS spectra during an insulin and glucose infusion for non-insulin-dependent diabetes mellitus subjects and healthy controls. The subjects with diabetes (solid symbols) synthesize glycogen slower than the control subjects (open symbols). Quantitation of the rates showed that insulin-stimulated glycogen synthesis is the major metabolic pathway of glucose disposal in both groups and that the impairment in muscle glycogen synthesis accounts for the reduced glucose disposal in the subjects with type 2 diabetes. (Reproduced from Shulman GI, Rothman DL, Jue T, Stein P, DeFronzo RA, Shulman RG, *New Engl J Med* **322**: 223–228, 1990 by permission of the Massachusetts Medical Society.)

4.2.2. Post-prandial Storage after a Mixed Meal in Healthy Subjects and Subjects with Type 2 Diabetes

The conditions of extreme hyperinsulinemia and hyperglycemia that occur in type 2 diabetes do not occur in healthy subjects under normal feeding conditions. In order to quantitate postprandial storage of glycogen in muscle under physiological conditions ^{13}C MRS has been used to measure the glycogen concentration of the gastrocnemius muscle in young healthy subjects over a 7 h period following a mixed meal of ordinary foodstuffs (3). With feeding there is a rapid rise in plasma glucose that initially outpaces the ability of the pancreas to release insulin. Several minutes after the onset of feeding the pancreas begins to release insulin in substantial quantities into the bloodstream, resulting in a fall of plasma glucose concentration to close to pre-meal values. In this study plasma glucose rose from 5.4 to 7.3 mmol in the first 30 min after a meal followed by a rapid fall to 6.4 mmol at 75 min and a progressive decrease to pre-meal levels at 240 min (29, 30). The normalization of plasma glucose concentration was associated with an increase in muscle glycogen concentration which started 1–2 h after ingestion of the meal. Muscle glycogen concentration rose from a mean of 83 to a peak of 100 mM at 4.9 h and fell thereafter to a mean of 91 mM at 7 h post-prandially. It was calculated from total muscle mass measurements and estimation of carbohydrate absorption rates that, at peak muscle glycogen concentrations, between 26 and 35 % of the total absorbed carbohydrate was stored as muscle glycogen. The approximately 2× lower fraction of glucose stored in muscle glycogen under these conditions compared with hyperglycemic hyperinsulinemia is most likely due to the lower glucose and insulin levels achieved physiologically in healthy subjects. However storage of glucose as muscle glycogen was still large enough to play an important role in post-prandial hyperglycemia if impaired.

The importance of the defect in muscle glycogen synthesis under conditions of a physiological meal was directly shown in a recent study by Carey *et al.* (4). In this study ^{13}C MRS was used to measure muscle glycogen concentrations before and after the consumption of sequential mixed meals (breakfast: 190.5 g carbohydrate, 41.0 g fat, 28.8 g protein, 1253 kcal; lunch: 203.3 g carbohydrate, 48.1 g fat, 44.0 g protein, 1497.5 kcal). Subjects with diet-controlled type 2 diabetes and age- and body mass index-matched non-diabetic controls were studied. After the first meal, mean glycogen concentration in the control group rose significantly from the basal level (from 69 to 97 mmol/l at 240 min). In the diabetic group, the post-prandial rise in glycogen was 3-fold lower than that of the control group (from 57 to 66 mmol/l at 240 min, $p < 0.005$,) despite considerably greater serum insulin levels (752.0 vs 372 pmol/l at 300 min, $p = 0.013$). This was associated with a significantly greater post-prandial hyperglycemia (10.8 vs 5.3 mmol/l at 240 min, $p < 0.005$). This study demonstrated directly that the impairment in insulin-stimulated muscle glycogen synthesis in type 2 diabetes plays a major role in post-prandial hyperglycemia under typical meal conditions.

4.3. IDENTIFICATION OF IMPAIRED ENZYMATIC STEPS IN INSULIN STIMULATED MUSCLE GLYCOGEN SYNTHESIS

The impairment in insulin-stimulated muscle glycogen synthesis in type 2 diabetes was anticipated by studies that found impairments in insulin-stimulated enzyme activities in the pathway. A schematic of the enzymatic steps in the muscle glycogen synthesis pathway is shown in Figure 4.2. Defects in glycogen synthase (5–8), hexokinase (9–11) and glucose transport (11–14) had all been implicated in the subnormal rate of muscle glycogen synthesis in type 2 diabetes. As described in Chapter 5, coordinate down-regulation of the activity of all of the pathway enzymes is consistent with the requirements for glucose-6-phosphate and glucose homeostasis. As described in Chapter 5, Metabolic Control Analysis was used (14) to show that GT/Hk controlled the flux of glycogen synthesis in healthy subjects, and that GSase served not to

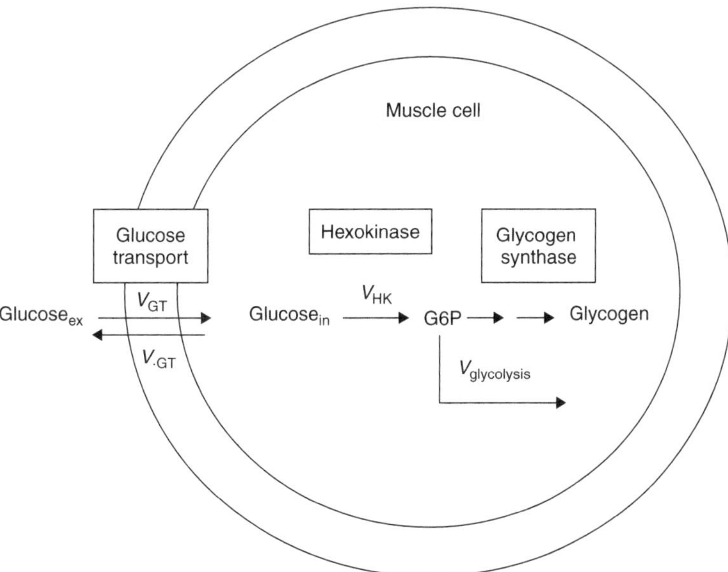

Figure 4.2. Diagram of potential impaired steps in insulin-stimulated muscle glycogen synthesis in type 2 diabetes. G6P, glucose- 6-phosphate; Glucose$_{ex}$, extracellular glucose; Glucose$_{in}$, intracellular glucose; $V_{glycolysis}$, net velocity of the glycolytic flux of glucose-6-phosphate; V_{GT}, velocity of glucose transport into the muscle cell; V_{-GT}, velocity of glucose transport out of the muscle cell; V_{HK}, velocity of glucose phosphorylation by hexokinase. (Reproduced from Shulman GI, Rothman DL, Jue T, Stein P, DeFronzo RA, Shulman RG, *New Engl J Med* **322**: 223–228, 1990 by permission of Massachusetts Medical Society.)

control flux but to maintain homeostasis of intermediates such as G6P. However in this chapter we will focus more on which of these steps plays the key role for determining the flux control of glycogen synthesis in type 2 diabetes and in insulin-resistant subjects with a genetic predisposition towards type 2 diabetes.

4.3.1. The Relative Importance of Impairments in Insulin-stimulated Muscle Glycogen Synthase Activity and Glucose Transport Activity in Type 2 Diabetes

The relative importance of these steps to insulin-stimulated muscle glycogen metabolism was assessed by performing $^{13}C/^{31}P$ MRS studies to measure intracellular concentrations of glucose, glucose-6-phosphate and glycogen in muscle of patients with type 2 diabetes and age–weight matched control subjects (11). Intracellular glucose-6-phosphate is an intermediary metabolite between glucose transport and glycogen synthesis, and hence the relative activities of these two steps will be reflected by its intracellular concentration. If the primary enzymatic defect in muscle glycogen synthesis was decreased activity of glycogen synthase in diabetes, glucose-6-phosphate in the diabetic patients would be expected be increased relative to that of the normal individuals. Using ^{31}P NMR to assess intracellular glucose-6-phosphate under conditions of hyperglycemic hyperinsulinemia, an increase of approximately 0.1 mM intracellular glucose-6-phosphate was observed in normal individuals, but no change in patients with type 2 diabetes. The blunted incremental change of glucose-6-phosphate concentration in the type 2 diabetic patients indicated that a selective defect in glycogen synthase was not responsible for their impaired glycogen synthesis. The elevated glucose-6-phosphate was consistent with the major activity difference responsible for the reduced rate being either decreased glucose transport activity or decreased hexokinase II activity.

4.3.2. Determination of Whether the Defect in Glucose Transport or Hexokinase II Activity in Type 2 Diabetes is Acquired or Genetic

To examine whether this defect in glucose transport or hexokinase II activity was a primary or acquired defect secondary to other factors, such as chronic hyperglycemia, insulin-resistant offspring of parents with type 2 diabetes were studied. Although these individuals were in all cases lean and normoglycemic, they are known to be at a ~40 % risk of developing diabetes, with an even higher risk in the most insulin-resistant offspring. The rate of muscle glycogen synthesis and the muscle glucose-6-phosphate concentration were measured under the same clamp conditions used to study subjects with type 2 diabetes (15). The children of diabetic parents exhibited a 50 % decrease in the rate of insulin-stimulated whole body glucose metabolism compared with age- and weight-matched control subjects, and this was mainly a consequence of decreased rates of muscle glycogen synthesis. They also showed a similar reduced insulin-stimulated increment of muscle glucose-6-phosphate, consistent with a similar impairment of glucose transport or hexokinase II activity to that seen in patients with fully developed type 2 diabetes.

In order to simulate the effect of a reduction in glucose transport or hexokinase II activity, control subjects were studied at similar insulin levels but at euglycemia. The 2× reduction in plasma glucose led to a 2× reduction in the rate of glycogen synthesis, similar to that measured in the offspring under hyperglycemia. The glucose-6-phosphate concentration also decreased to values similar to that of the type 2 diabetic offspring. These results were again highly consistent with reduced insulin-stimulated activity in glucose transport or hexokinase II activity being the primary factor in the reduced insulin stimulated glycogen synthesis in the offspring of parents with type 2 diabetes. Therefore, even prior to the onset of diabetes, insulin-resistant offspring of patients with type 2 diabetes have decreased rates of muscle glycogen synthesis that are secondary to a defect in muscle glucose transport or hexokinase activity. Furthermore, from these results, summarized in Figure 4.3 it may be concluded that the reduction in the insulin stimulated

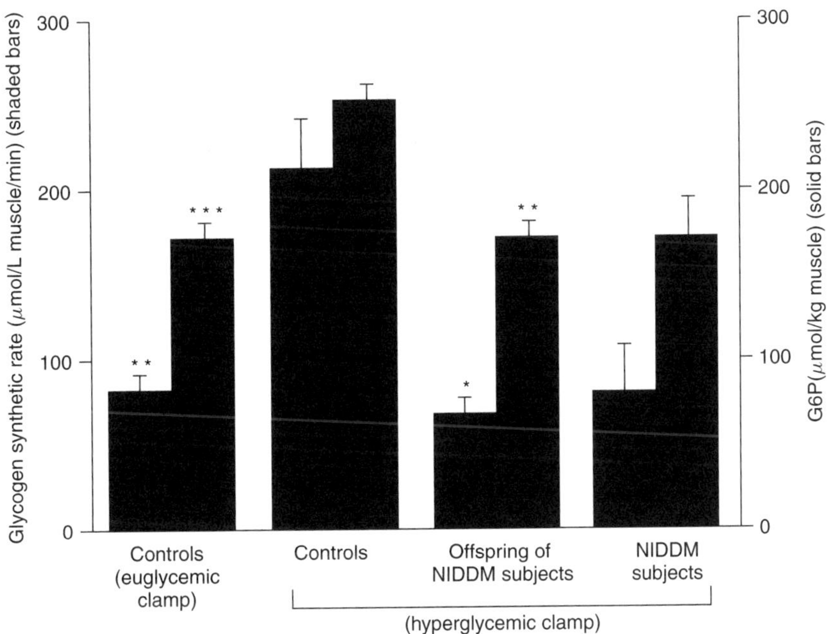

Figure 4.3. Mean rates of muscle glycogen synthesis and muscle glucose-6-phosphate (G6P) concentration in the offspring of parents with non-insulin-dependent (type 2) diabetes mellitus (NIDDM), control subjects and subjects with NIDDM during the same conditions of hyperglycemia–hyperinsulinemia and in the same control subjects under conditions of euglycemia–hyperinsulinemia. Both the mean rate of glycogen synthesis and glucose-6-phosphate concentration were reduced in the offspring of parents with NIDDM to values similar to these previously observed in subjects with NIDDM. $*p < 0.005$, $**p < 0.003$; $***p < 0.0004$ (compared with the control subjects at hyperglycemic hyperinsulinemia). (Reproduced from Rothman DL, Magnusson I, Cline CG, Gerard D, Kahn CR, Shulman RG, Shulman GI, *Proc. Natl Acad Sci. USA* **92**: 983–987, 1995 by permission of the National Academy of Sciences.)

activity of enzymes in the muscle glycogen synthesis pathway occurs early in the pathogenesis of type 2 diabetes.

4.3.3. Determination of Whether Glucose Transport or Hexokinase II is Rate-controlling for Insulin-stimulated Muscle Glycogen Synthesis in Type 2 Diabetes

Because the activities determined from biopsy samples and cellular systems of both glucose transport and hexokinase II activities are reduced in models of type 2 diabetes, the relative importance of these two reactions has been controversial. As described in Chapter 3, the step with the major impact on the rate will have the highest flux control coefficient if the reduction of total activity is similar. To determine whether glucose transport or hexokinase II activity is rate-controlling for insulin-stimulated muscle glycogen synthesis in patients with type 2 diabetes, a novel ^{13}C NMR method was developed using a combined infusion of $[1-^{13}C]$ glucose and $[2-^{13}C]$ mannitol. The mannitol is not taken up by muscle, and therefore its signal is proportional to the extracellular plus vascular space in the MRS voxel. By measurement of the ratio of plasma glucose and mannitol the contribution of extracellular glucose to the MRS spectrum can be calculated. This method was applied to assess intracellular glucose concentrations in muscle under similar hyperglycemic–hyperinsulinemic conditions as used in the previous MRS studies (16). Intracellular glucose is an intermediate between glucose transport and glucose phosphorylation, and its concentration reflects the relative activities of glucose transporters and of hexokinase II. Unlike the standard biopsy method, this approach is non-invasive and is not subject to the errors caused by contamination of biopsy tissue with plasma glucose or incomplete removal from the biopsy of non-muscle constituents. Using similar reasoning as in the study of glucose-6-phosphate, if hexokinase II activity was decreased in diabetes relative to glucose transport activity, a substantial increase in intracellular glucose would be predicted, whereas if glucose transport was primarily responsible for maintaining intracellular glucose metabolism, intracellular glucose and glucose-6-phosphate should change proportionately. The experimental measurements showed that intracellular glucose concentration was lower in the subjects with type 2 diabetes than in control subjects. In both control and diabetic subjects intracellular glucose was far lower than the value expected if hexokinase II was the primary rate-controlling enzyme for glycogen synthesis (16).

4.4. DETERMINATION OF THE METABOLIC MECHANISMS THROUGH WHICH OBESITY AND EXERCISE AFFECT INSULIN SENSITIVE MUSCLE GLYCOGEN SYNTHESIS

The studies of the offspring of patients with type 2 diabetes implied that genetics play a primary role in the development of insulin resistance. However this finding did not explain environmental factors, particularly low physical activity and high-calorie high-fat diets leading to elevated plasma free fatty acid levels and obesity. The question arose whether genetic and environmentally induced insulin insensitivity were both mediated through glucose transport or were environmental effects mediated through other metabolic steps? In this section MRS studies are reviewed that have examined the metabolic mechanism by which plasma free fatty acid concentrations and exercise influence insulin-stimulated muscle glycogen synthesis in non-insulin-resistant subjects. It is shown that in human muscle exposure to free fatty acids induces muscle insulin resistance via down-regulation of glucose transport activity. This mechanism differs from the classic mechanism proposed by Randall in which fatty acid oxidation inhibits glycolysis via the effect of citrate on phosphofructokinase (PFK). In both insulin-resistant offspring and insulin-sensitive controls physiological exercise results in an up-regulation of insulin-stimulated glucose transport and glucose storage as glycogen. Taken as a whole, these findings suggest that the major difference in the insulin-resistant offspring is a greater sensitivity to environmental effects, as opposed to a difference in the mechanism by which environmental factors influence insulin-stimulated glycogen synthesis.

4.4.1. $^{13}C/^{31}P$ MRS Studies Find that Free Fatty Acid-induced Muscle Insulin Resistance is Mediated via a Reduction in Glucose Transport/Phosphorylation Activity

Increased plasma free fatty acid concentrations are typically associated with many insulin-resistant states, including obesity and type 2 diabetes mellitus (21–24). Increased plasma free fatty acid concentrations are typically associated with many insulin-related states, including obesity and type 2 diabetes mellitus (17–24). An inverse relationship between fasting plasma fatty acid concentrations and insulin sensitivity was found in a cross-sectional study of young, normal-weight offspring of type 2 diabetic patients, consistent with the hypothesis that altered fatty acid metabolism contributes to insulin resistance in patients with type 2 diabetes (25). The findings of a relationship between free fatty acids and insulin resistance has been extended by recent studies measuring intramuscular triglyceride content by muscle biopsy (21, 22) or intramyocellular triglyceride content by ^{1}H NMR (23–27). These studies have shown an even stronger relationship between accumulation of intramyocellular triglyceride and insulin resistance.

The mechanism originally proposed by Randle to explain how, in the heart, fatty acid oxidation can inhibit glucose metabolism was via alterations in phosphofructokinase activity due to an increase in mitochondrial oxidation of fatty acids. In this model increased fatty acid oxidation leads to an elevation of the intramitochondrial acetyl-CoA:CoA and NADH:NAD^+ ratios, with subsequent inactivation of pyruvate dehydrogenase (29, 30). The elevation of the NADH:NAD^+ ratio in turn causes an increase in intracellular citrate concentration, leading to inhibition of phosphofructokinase, believed to be a key rate-controlling enzyme in glycolysis. Subsequent accumulation of glucose-6-phosphate would inhibit hexokinase II activity, resulting in a decrease in glycolysis and glycogen synthesis.

A series of recent studies using MRS have challenged the Randle hypothesis in the case of muscle metabolism under high insulin conditions (31, 32). In the initial study, ^{13}C and ^{31}P MRS was used to measure skeletal muscle glycogen and glucose-6-phosphate concentrations in healthy subjects. The subjects were maintained in euglycemic hyperinsulinemic conditions with either low or high levels of plasma fatty acids (31, 32). Increasing the plasma fatty acid concentration for 5 h caused a $\sim$50 % reduction in rates of insulin-stimulated rates of muscle glycogen synthesis and whole body glucose oxidation compared with the control studies. In contrast to the predictions of the Randle model, where fat-induced insulin resistance should result in an increase in intramuscular glucose-6-phosphate due to inhibition of PFK, it was observed that the fall in muscle glycogen synthesis was preceded by a decrease in the intramuscular muscle glucose-6-phosphate content. These data suggest that increases in plasma fatty acid concentrations induce insulin resistance by leading to a reduction in glucose transport or phosphorylation activity, from which a reduction in muscle glycogen synthesis and glucose oxidation follows. A time course of glucose-6-phosphate concentration, muscle glycogen concentration prior to and during the lipid infusion is shown in Figure 4.4.

The reduction in insulin-activated glucose transport/phosphorylation activity in normal subjects maintained at high plasma fatty acid levels was similar to that seen in obese individuals (33), patients with type 2 diabetes (11), and lean normoglycemic insulin-resistant offspring of type 2 diabetic individuals (15). Hence, accumulation of intramuscular fatty acids (or fatty acid metabolites) appears to play an important role in the pathogenesis of insulin resistance seen in obese patients and patients with type 2 diabetes. Moreover, fatty acids would seem to interfere with a very early step in insulin-stimulation of Glut-4 transporter activity or hexokinase II activity.

4.4.2. Measurement of Intracellular Glucose Indicates that free fatty acids Reduce Insulin-stimulated Muscle Glycogen Synthesis Primarily through Reduction of Glucose Transport Activity

In order to distinguish whether free fatty acids primarily influence the rate of insulin-stimulated muscle glucose uptake via reduction of glucose transport activity or on hexokinase II activity, intracellular concentrations of glucose were measured in muscle using ^{13}C NMR (32). As described above for glucose-6-phosphate, because intracellular glucose is an intermediate between glucose transport and hexokinase II, its concentration reflects the relative activities of these two steps. If a selective decrease in hexokinase activity was responsible for the lower rate of insulin-stimulated muscle glycogen synthesis, intracellular glucose concentrations should increase until the reduced phosphorylation of glucose is matched by reverse transport of glucose out of the muscle. However, if the reduced glucose transport activity exerts primary flux control there should be no difference or a decrease in the intracellular glucose concentration. Elevated plasma fatty acid concentrations caused a significant reduction in intracellular glucose concentration in the lipid infusion studies compared with control studies in which glycerol (the other metabolite released by lipolysis) was infused in the absence of any exogenous fatty acid. Furthermore intracellular glucose concentrations were exceedingly low, consistent with minimal glucose transporter back flux and control of

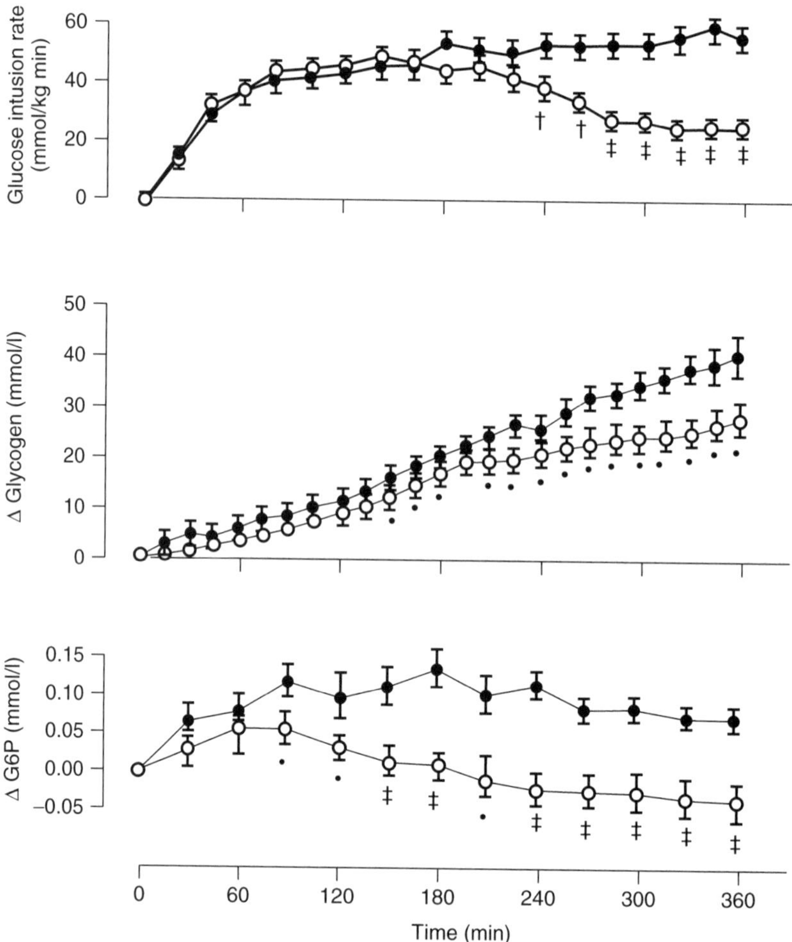

Figure 4.4. Effect of increasing plasma fatty acid in nine healthy subjects. Glucose infusion rate, increase in calf muscle glycogen, and increase in calf muscle glucose-6-phosphate (G6P) at low (solid circles) and elevated (open circles) fatty acid concentrations. $^*p < 0.05$; $^\dagger p < 0.01$; $^\ddagger p < 0.001$. (Reproduced from Dresner A, Laurent D, Marcucci M, Griffin ME, Dufour S, Cline GW, Slezak LA, Andersen DK, Hundal RS, Rothman DL, Petersen KF, Shulman GI, *J. Clin. Invest.* **103**: 253–259, 1999 by permission of the American Society for Clinical Investigations.).

flux at the transport step. These data imply that the primary rate-controlling step for fatty acid induced impaired insulin-stimulated glycogen synthesis in humans is glucose transport, similar to what was found in type 2 diabetes and insulin-resistant offspring of patients with diabetes.

4.4.3. Exercise Enhances Insulin-stimulated Glycogen Synthesis by Up-regulation of Glucose Transport/Phosphorylation Activity

Physical activity has been known for several decades to improve insulin sensitivity. To assess whether regular exercise improved insulin sensitivity through an enhancement of insulin-stimulated muscle glycogen synthesis young, healthy, insulin-resistant offspring of non-insulin-dependent diabetes mellitus parents and controls participated in 6 weeks of moderate aerobic training, consisting of 45 min exercise every other day. Regular exercise increased the muscle glucose-6-phosphate concentration and rate of muscle glycogen synthesis during a hyperglycemic clamp of the offspring to the same values as measured in the control subjects prior to the exercise regimen. Studies were performed the day after the previous exercise bout to rule out potential effects of exercise-stimulated muscle glucose transport. Regular exercise also increased insulin sensitivity, thus demonstrating that this abnormality can be reversed with exercise training (34). However control subjects also substantially increased their rate of insulin-stimulated glycogen synthesis in the study. The reversal of the defect through regular exercise in these subjects underlines the complex interactions between the environment (lifestyle) and genetics in the development of the disease.

4.5. SUMMARY AND CONCLUSION

MRS has provided a powerful tool for the study of the pathogenesis and etiology of insulin resistance. Prior to the studies described in this chapter, the importance of muscle glycogen synthesis in whole body insulin resistance was controversial, and any defect in this pathway was attributed to reduced glycogen synthase activity. The initial studies using ^{13}C MRS showed that impaired insulin-stimulated muscle glycogen synthesis is in fact responsible for most of the insulin resistance observed in skeletal muscle of patients with type 2 diabetes, and is the major factor in the reduced whole body insulin sensitivity in these patients. Combined ^{13}C and ^{31}P MRS studies then showed that this abnormality is primarily due to reduced insulin-stimulated muscle glucose transport activity, as opposed to glycogen synthase, consistent with the high flux control coefficient of this step described in Chapter 5. A similar metabolic abnormality was found in healthy lean insulin-resistant offspring of patients with type 2 diabetes, indicating that this defect occurs early in the pathogenesis of type 2 diabetes. These abnormalities in insulin sensitivity are strongly associated with increased intramyocellular lipid accumulation as found by ^{1}H MRS (22–28). The mechanism by which elevated free fatty acids reduce insulin-stimulated muscle glycogen synthesis was shown using MRS to differ from the classic mechanism of fat-induced insulin resistance suggested by Randle *et al.* (29, 30). Instead the MRS studies found that elevated free fatty acids lead to a down-regulation of insulin-stimulated glucose transport, to activity levels similar to those observed in type 2 diabetes and in inherited insulin resistance. These studies strongly implicate altered intramyocellular lipid metabolism as having a major role in the pathogenesis of insulin resistance through interaction with glycogen metabolism, most likely via insulin signaling pathways. The ability provided by MRS to study the quantitative effects of environmental and genetic factors on insulin-stimulated glycogen metabolism provides a strong foundation for ongoing investigations targeted at a comprehensive understanding of insulin resistance at the molecular level.

Acknowledgments

This work was supported by grants R01 AG-23686, R01 DK- 063192, R01 DK-49230, P30 DK-45735 and M01 RR-00125. We also acknowledge ongoing support from the Yale General Clinical Research Center.

REFERENCES

1. Ferrannini E, Smith JD, Cobelli C, Toffolo G, Pilo A, DeFronzo RA. Effect of insulin on the distribution and disposition of glucose in man. *J Clin Invest* **76**(1): 357–364, 1985.
2. Shulman GI, Rothman DL, Jue T, Stein P, DeFronzo RA, Shulman RG. Quantitation of muscle glycogen synthesis in normal subjects and subjects with non-insulin dependent diabetes by ^{13}C nuclear magnetic resonance spectroscopy. *New Engl J Med* **322**: 223–228, 1990.
3. Taylor R, Price TB, Katz LD, Shulman RG, Shulman GI. Direct measurement of change in muscle glycogen concentration after a mixed meal in normal subjects. *Am J Physiol* **265**: E224–229, 1993.

4 Carey PE, Halliday J, Snaar JE, Morris PG, Taylor R. Direct assessment of muscle glycogen storage after mixed meals in normal and type 2 diabetic subjects. *Am J Physiol Endocrinol Metab* **284**(4): E688–694, 2003.
5. Bogardus C, Lillioja S, Stone K, Mott D. Correlation between muscle glycogen synthase activity and *in vivo* insulin action in man. *J Clin Invest* **73**: 1185–1190, 1984.
6. Johnson AB, Argyraki M, Thow JC, Jones IR, Broughton D, Miller M, Taylor R. Impaired activation of muscle glycogen synthase in non-insulin dependent diabetes mellitus is unrelated to the degree of obesity. *Metabolism* **40**: 252–260, 1991.
7. Johnson AB, Argyraki M, Thow JC, Jones IR, Broughton DL, Miller M, Taylor R. The effect of sulphonylurea therapy on skeletal muscle glycogen synthase activity and insulin secretion in newly presenting non-insulin dependent diabetic patients. *Diabet Med* **8**: 243–253, 1991.
8. Wright KS, Beck-Nielsen H, Kolterman OG, Mandarino LJ. Decreased activation of skeletal muscle glycogen synthase by mixed meal ingestion in NIDDM. *Diabetes* **37**: 436–440, 1988.
9. Braithewaite SS, Plazuk B, Colca JR, Edwards CW, Hofmann C. Reduced expression of hexokinase II in insulin resistant diabetes. *Diabetes* **44**: 43–48, 1995.
10. Kruszynska YT, Mulford MI, Baloga J, Yu JG, Olefsky JM. Regulation of skeletal muscle hexokinase II by insulin in nondiabetic and NIDDM subjects. *Diabetes* **47**: 1107–1113, 1998.
11. Rothman DL, Shulman RG, Shulman GI. ^{31}P nuclear magnetic resonance measurements of muscle glucose-6-phosphate. Evidence for reduced insulin-dependent muscle glucose transport or phosphorylation activity in non-insulin-dependent diabetes mellitus. *J Clin Invest* **89**: 1069–1075, 1992.
12. Bonadonna RC, Del Prato S, Bonora E. Roles of glucose transport and glucose phosphorylation in muscle insulin resistance of NIDDM. *Diabetes* **45**: 915–925, 1996.
13. Dohm GL, Tapscott EB, Pories WJ. An *in vitro* human preparation suitable for metabolic studies. Decreased insulin stimulation of glucose transport in muscle from morbidly obese and diabetic subjects. *J Clin Invest* **82**: 486–494, 1988.
14. Shulman RG, Rothman DL. Enzymatic phosphorylation of muscle glycogen synthase: A mechanism for maintenance of metabolic homeostasis. *Proc. Natl. Acad. Sci. USA* **93**: 7491–7495, 1996.
15. Rothman DL, Magnusson I, Cline G, Gerard D, Kahn CR, Shulman RG, Shulman GI. Decreased muscle glucose transport/phosphorylation is an early defect in the pathogenesis of non-insulin-dependent diabetes mellitus. *Proc Natl Acad Sci USA* **92**: 983–987, 1995.
16. Cline GW, Petersen KF, Krssak M, Shen J, Hundal RS, Trajanoski Z, Inzucchi S, Dresner A, Rothman DL, Shulman GI. Decreased glucose transport as a cause of decreased insulin-stimulated muscle glycogen synthesis in Type 2 diabetes. *New Engl J Med* **341**: 240–246, 1999.
17. Frayn KN. Insulin resistance and lipid metabolism. *Curr Opin Lipidol* **1**: 197–204, 1993.
18. Boden G, Chen X, Ruiz J, White JV, Rossetti L. Mechanism of fatty acid induced inhibition of glucose uptake. *J Clin Invest* **93**: 2438–2446, 1994.
19. Argyraki M, Wright PD, Venables CW, Proud G, Taylor R. Study of human skeletal muscle *in vitro*: effect of NEFA supply on glucose storage. *Metabolism* **38**: 1183–1187, 1989.
20. Johnson AB, Argyraki M, Thow JC, Cooper BG, Fulcher G, Taylor R. Effect of increased free fatty acid supply on glucose metabolism and skeletal muscle glycogen synthase activity in normal man. *Clin Sci* **82**: 219–226, 1992.
21. Phillips DIW, Caddy S, Ilic V, Fielding BA, Frayn K, Borthwick AC, Taylor R. Intramuscular triglyceride and muscle insulin sensitivity: evidence for a relationship in nondiabetic subjects. *Metabolism* **45**: 947–950, 1996.
22. Pan DA, Lillioja S, Kridetos AD. Skeletal muscle triglyceride levels are inversely related to insulin action. *Diabetes* **46**: 983–988, 1997.
23. Krssak M, Falk Petersen K, Dresner A, DiPietro L, Vogel SM, Rothman DL, Shulman GI, Roden M. Intramyocellular lipid concentrations are correlated with insulin sensitivity in humans: a ^{1}H NMR spectroscopy study. *Diabetologia* **42**: 113–116, 1999.
24. Falholt K, Jensen I, Jensen SL, Mortensen H, Volund A, Heding LG, Petersen P, Falholt W. Carbohydrate and lipid metabolism of skeletal muscle in type 2 diabetic patients. *Diabet Med* **5**: 27–31, 1988.
25. Perseghin G, Ghosh S, Gerow K, Shulman GI. Metabolic defects in lean nondiabetic offspring of NIDDM parents: a cross-sectional study. *Diabetes* **46**: 1001–1009, 1997.

26. Szcepaniak LS, Babcock EE, Schick F, Dobbins RL, Garg A, Burns DK, McGarry JD, Stein DT. Measurement of intracellular triglyceride levels by H spectroscopy: validation *in vivo*. *Am J Physiol* **276**: E977–E989, 1999.
27. Jacob S, Machann J, Rett K, Brechtel K, Volk A, Renn W, Maerker E, Matthaei S, Schick F, Claussen CD, Haring HU. Association of increased intramyocellular lipid content with insulin resistance in lean nondiabetic offspring of type 2 diabetic subjects. *Diabetes* **48**: 1113–1119, 1999.
28. Perseghin G, Scifo P, De CF, Pagliato E, Battezzati A, Arcelloni C, Vanzulli A, Testolin G, Pozza G, Del MA, Luzi L. Intramyocellular triglyceride content is a determinant of *in vivo* insulin resistance in humans: a ^{1}H–^{13}C nuclear magnetic resonance spectroscopy assessment in offspring of type 2 diabetic parents. *Diabetes* **48**: 1600–1606, 1999.
29. Randle PJ, Garland PB, Hales CN, Newsholme EA. The glucose fatty acid cycle its role in insulin sensitivity and the metabolic disturbances of diabetes mellitus. *Lancet* **i**: 785–789, 1963.
30. Randle PJ, Kerbey AL, Espinal J. Mechanisms decreasing glucose oxidation in diabetes and starvation: role of lipid fuels and hormones. *Diabetes Metab Rev* **4**: 623–638, 1988.
31. Roden M, Price TB, Perseghin G, Petersen KF, Rothman DL, Cline GW, Shulman GI. Mechanism of free fatty acid-induced insulin resistance in humans. *J Clin Invest* **97**: 2859–2865, 1996.
32. Dresner A, Laurent D, Marcucci M, Griffin ME, Dufour S, Cline GW, Slezak LA, Andersen DK, Hundal RS, Rothman DL, Petersen KF, Shulman GI. Effects of free fatty acids on glucose transport and IRS-1-associated phosphatidylinositol 3-kinase activity. *J Clin Invest* **103**: 253–259, 1999.
33. Petersen KF, Hendler R, Price T, Perseghin G, Rothman DL, Held N, Amatruda JM, Shulman GI. $^{13}C/^{31}P$ NMR studies on the mechanism of insulin resistance in obesity. *Diabetes* **47**: 381–386, 1998.
34. Perseghin G, Price TB, Petersen KF, Roden M, Cline GW, Gerow K, Rothman DL, Shulman GI. Increased glucose transport – phosphorylation and muscle glycogen synthesis after exercise training in insulin resistant subjects. *New Engl J Med* **335**: 1357–1362 (1996).

5

Phosphorylation of Allosteric Enzymes can serve Homeostasis rather than Control Flux: the Example of Glycogen Synthase

James R.A. Schafer

Department of Neuroscience, Yale University School of Medicine, MR Center, PO Box 208043, New Haven, CT 06520-8043, USA

David A. Fell

School of Biological and Molecular Sciences, Oxford Brookes University, Gipsy Lane, Oxford OX3 OBP, UK

Douglas L. Rothman

Department of Diagnostic Radiology, Yale University School of Medicine, MR Center, PO Box 208043, New Haven, CT 06520-8043, USA

Robert G. Shulman

Department of Diagnostic Radiology, Yale University School of Medicine, MR Center, PO Box 208024, New Haven, CT 06520-8024, USA

Metabolomics by In Vivo NMR. Edited by R. G. Shulman and D. L. Rothman
 ISBN: 0-470-84719-0

5.1. INTRODUCTION

Metabolic control analysis (MCA) has revolutionized our understanding of flux control through theoretical and experimental studies (1–3). One of its key precepts is the flux control coefficient, a quantitative characterization of the influence of any enzyme over the flux of a metabolic pathway. As MCA has developed, it has expanded into associated areas. One of particular interest has been the mechanism of metabolite control – in which the central question is how metabolite homeostasis, the constancy of metabolite concentrations, is maintained during changes in flux (4–8). One well-known example of such metabolite homeostasis comes from the study of fluxes through the energetic pathway of glucose oxidation. There, the flux can change by more than an order of magnitude while metabolites remain constant within small experimental error (9, 10). While theoretical analyses of such homeostatic phenomena have been available since the first description of MCA, physical explorations of this important physiological area have been scarce because of experimental limitations. The primary hurdle has been the difficulties in simultaneously obtaining *in vivo* values of metabolite concentrations and fluxes through the pathway.

A central theme of this book is how these limitations have been somewhat eased by the development of *in vivo* NMR methods. Valuable information about energetics has been obtained by ^{31}P NMR of high-energy metabolites (e.g. ATP, PO_4 and creatine phosphate) and about metabolic regulation from phosphorylated pathway intermediates such as glucose-6-phosphate (G6P) (11). As shown elsewhere in this book, the weak natural abundance (1.1 %) of the ^{13}C isotope allows direct ^{13}C NMR of labeled ^{13}C substrates to measure flux, by following the time course of label flow into metabolic pools (12). A series of studies in which ^{31}P and ^{13}C NMR are applied to one pathway, together with the evaluation of enzyme parameters required by MCA, has enabled us to obtain a combined understanding of flux control and homeostasis. These results are presented in this chapter for a particular pathway – that of glycogen synthesis in the rat gastrocnemius muscle. Since enzymatic phosphorylation and allostery have dominant roles in the control of this pathway, this two-fold experimental approach has revealed conclusions of considerable generality about the functions served *in vivo* by enzymatic phosphorylation.

The analysis of control and regulation of metabolism in terms of supply and demand blocks by Hofmeyr and Cornish-Bowden (4, 5) is relevant here, since they showed how homeostasis of an intermediate in the pathway can be framed in terms of the supply of, and demand for, this metabolite. The degree of homeostasis of a metabolite that can be expected when an external stimulus changes flux in one of the blocks depends upon the direct kinetic effects of the intermediate on both blocks through substrate, product and effector interactions. Allostery helps to achieve metabolite homeostasis by accentuating the response to an intermediate but is limited by the cooperativity of the allosteric response.

It has been known that better homeostasis of metabolites could be obtained if flux changes were brought about by stimuli acting at more than one point. Kacser and Acerenza had (13) suggested that fluxes could be increased while maintaining intermediate constancy by simultaneously increasing the activity of enzymes both immediately up- and downstream from each metabolite in exact proportion. They posited that without such alterations intermediate metabolite concentrations would change and could perturb linked pathways. Since multiple enzymes must be altered simultaneously to fulfill this postulate of homeostasis, gene regulation was suggested as a useful mechanism because shared promoters can link the concentrations of different proteins and changes can be of any duration. However, transcription and translation can take minutes or even hours to exert their full effect. While many examples are known where chronic changes in metabolic state are accompanied by similar coordinated changes of all the enzymes in the pathway (14), an organism depending solely on genetic regulation to respond to metabolic changes, which can change on the seconds time scale or even faster, would find itself perpetually behind the times. To respond more effectively, it would need more dynamic systems.

It has been proposed that selected multi-site post-translational modulation, in particular enzymatic phosphorylation, could substitute for the cumbersome genetic changes in enzymatic concentrations (3, 6, 15). Such multi-site modulation could retain regulation of metabolites during large-scale changes in enzymatic concentration by changing the activity of selected enzymes. These multi-site control points would change activity in order to exercise metabolite control. In a specific example, glycogen synthase (GSase), an allosteric and reversibly phosphorylated enzyme that was once thought of as flux-controlling, has been proposed to regulate metabolite concentration (15).

To test the hypothesis that modulation by both allosteric and phosphorylation mechanisms might serve this homeostatic role, we extended the earlier studies of the role of muscle GSase in flux control and G6P homeostasis (15, 16). This enzyme has at least 12 phosphorylation sites and is sensitive to some half-dozen allosteric effectors – most notably the upstream metabolite, G6P (17). Additionally, it was long thought of as the rate-limiting enzyme in glycogen synthesis (17). However, the flux control of this pathway in mammalian muscle [whose derangement in non-insulin-dependent diabetes mellitus (NIDDM) appears to be one of the major causes of pathology] has been shown to reside predominantly in the glucose transporter (Figure 5.1) (15–16, 18–21). Furthermore three-fold pharmacological activation of muscle GSase activity has no effect on insulin-stimulated muscle glycogen synthesis (22). Similarly, in rat hepatocytes, control of glycogen synthesis is shown to reside in the hexokinase IV, or glucokinase (23), as opposed to glycogen synthase. These findings highlighted the question of why, given its small role in flux control in both tissues, GSase velocity needs so much regulation. Using new results generated from *in vivo* NMR measurements in our laboratory, as well as older data from Rossetti and Hu (24), we now show that the purpose of the regulation of GSase by phosphorylation and by allostery is to maintain steady concentrations of pathway intermediates despite huge changes in pathway flux.

5.2. METHODS

5.2.1. Metabolic Control Analysis

MCA gets its strength from deriving consequences from quantitative definitions of parameters. Subsequent developments of MCA theory have been devoted to facilitating these parameters by available experimental

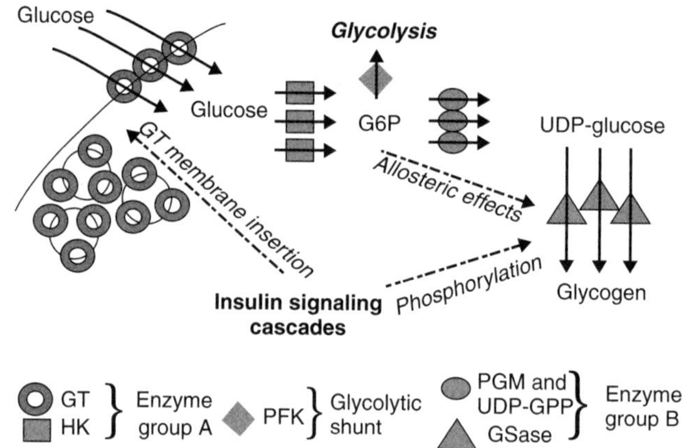

Figure 5.1. Muscle glycogen synthetic pathway. Glucose transporters (GT), divided between an active membrane fraction and an inactive vesicle fraction, move plasma glucose into the cytoplasm where it is phosphorylated by hexokinase (HK) into glucose-6-phosphate (G6P). Isomerized G6P can serve as the substrate for phosphofructo-kinase (PFK), the entry point into glycolysis. However, under conditions favoring glycogen synthesis a majority of G6P is converted by phosphoglucomutase (PGM) and UDP-glucose pyrophosphorylase (UDP-GPP) into UDP-glucose. UDP-glucose is the substrate for glycogen synthase (GSase), which is ultimately responsible for making glycogen. The pathway is regulated by downstream insulin-induced effects in at least two locations. One, insulin is the primary determinant of the distribution of GT between membrane and vesicles. Two, insulin affects GSase velocity and allosteric responsiveness via phosphorylation. Additionally, G6P can induce its own consumption by allosterically stimulating GSase. The bottom section of the drawing indicates the distribution of the glycogen synthetic enzymes into subsystems A and B.

methods. *In vivo* MRS has extended these empirical evaluations. These parameters, e.g. flux control coefficients elasticities and responsivities, have been described in Chapter 3 and are briefly reviewed below, because the evaluation of homeostasis follows directly from their experimental evaluation. The strengths of MCA are difficult to overstate since the starting assumptions and definitions lead unequivocally to the conclusions, so that, in the present determination of the regulation of homeostasis in the particular pathway of glycogen synthesis, once the data are available the conclusions drawn by MCA are simply consequences. This introduces a higher degree of definiteness than is available in other analyses of metabolism so that the description of homeostasis now discussed is simply a deduction from the definitions used in MCA and the data upon which the conclusions are drawn.

As discussed in Chapter 3, the control of an enzyme, E, over pathway flux can be expressed as a flux control coefficient C_E^J, which is defined as the fractional change in pathway flux (J) over the fractional change in enzyme activity:

$$C_E^J = (\partial J/J)/(\partial E/E) = \partial \ln J/\partial \ln E \tag{5.1}$$

The elasticity of an enzyme with respect to a given metabolite, ε_M^E, is expressed as the fractional change in the activity of that enzyme (E) for a given fractional change in metabolite concentration (M):

$$\varepsilon_M^E = (\partial E/E)/(\partial M/M) = \partial \ln E/\partial \ln M \tag{5.2}$$

The connection between elasticities and the flux control coefficients is given by the connectivity relationship (1):

$$\sum_{E=1}^{n} C_E^J \varepsilon_M^E = 0 \tag{5.3}$$

This states that for any metabolite M, the product of the flux control coefficients of the n enzymes with their elasticities with respect to M sums to zero. As a result, where only the two enzymes producing and consuming M respond to it, there is a tendency for the enzyme with a lower value of elasticity compared with the other to have a high control coefficient and vice versa. This condition applies in the supply–demand analysis below.

While the above terms and relationships were initially defined for single enzymes, they have been shown to be equally valid for groups of contiguous steps in a pathway provided certain conditions are met (25). The most extreme case of this simplification, in which long pathways can be considered as two subsystems separated by a common metabolite, is called 'top-down' analysis (26). In the pathway from glucose to glycogen, provided the branch flux to glycolysis is relatively small as it is in resting muscle, the two subsystems can be defined as: (A) the GT/HK subsystem that generates G6P from blood glucose, and (B) the GSase subsystem that converts G6P into glycogen (Figure 5.1).

5.2.2. Supply–Demand Analysis

In a supply–demand system such as the one treated here, as stated above, the distribution of flux control depends on the ratio of the supply and demand elasticities to G6P: the lower the ratio the more control shifts to the supply. On the other hand, the degree of homeostatic maintenance of this linking metabolite, when either the supply or the demand block is activated, depends on the G6P concentration control coefficients of the two blocks, which are inversely proportional to the sum of the supply and demand elasticities (3–5). Hence, the greater this sum, the smaller the impact of activation on G6P concentration, and the better the homeostasis. If one reaction block controls the flux, then the maintenance of homeostasis of the linking metabolite becomes the function of the other reaction block. If all of the flux control is, say, in the supply (which implies that the demand elasticity $\gg$ supply elasticity), then, in this instance, homeostasis depends only on the demand elasticity (5).

5.2.3. General Proportionality Analysis

A proportional activation term (π) quantitatively compares the effect of some external factor (X) upon the activities of subsystems A and B (7). The most general expression of π is in terms of the relative elasticities of A and B to X:

$$\pi_X^{AB} = \frac{\varepsilon_X^B}{\varepsilon_X^A} = \left(\frac{\partial B/B}{\partial A/A}\right)_X \tag{5.4}$$

Qualitatively, Equation (5.4) indicates that, if X acts primarily through subsystem B then π_X^{AB} values will be >1, and if it acts through subsystem A then π_X^{AB} values will be <1. π_X^{AB} values of unity indicate equivalent action of X on both A and B. Negative values, which would indicate that X stimulates one subsystem while suppressing the other, were not considered here – although they may be applicable to other pathways.

The response of the system to a change in X will contain contributions from the direct changes in activity of blocks A and B induced by X and from the changes in activity of the blocks caused by the consequential changes in the linking metabolite, M, between the blocks. Korzeniewski *et al.* (7) showed that, by taking these components of the response into account, Equation (5.4) could be transformed into:

$$\pi_X^{AB} = \frac{1 - \varepsilon_M^B[(\partial \ln M)_X/(\partial \ln J)_X]}{1 - \varepsilon_M^A[(\partial \ln M)_X/(\partial \ln J)_X]} \tag{5.5}$$

The term $(\partial \ln M/\partial \ln J)$ that appears in the numerator and denominator can be obtained as the slope of a graph of $\ln M$ vs $\ln J$ for various values of X, and is also known as $\Omega_X^{M:J}$, the co-response coefficient of

M and J to X (27), or τ (7). Rewriting Equation (5.5) in terms of τ gives:

$$\pi_X^{AB} = (1 - \varepsilon_M^B * \tau)/(1 - \varepsilon_M^A * \tau) \tag{5.6}$$

This form provides boundary conditions for the description of π_X^{AB}. Substituting $\tau = 1/\varepsilon_M^B$ into Equation (5.6) shows π_X^{AB} is then zero, so, from the definition of π_X^{AB}, the effector must affect flux entirely via subsystem A and any changes in subsystem B would be the result of feed-forward stimulation via the connecting metabolite (in this case G6P). Similarly, if τ approaches $1/\varepsilon_M^A$ then π_X^{AB} goes toward infinity and the effector operates in the opposite manner – A is 'pulled' along by a drop in M (in this case G6P). Finally, if M is unchanged when J is stimulated then $\tau = 0$, $\pi_X^{AB} = 1$, the two subsystems have been activated an equal amount with respect to X, and the metabolite is not playing a direct role in activating either subsystem. In this last case, the teleologically inclined might state that X is modulating both A and B *for the purpose* of maintaining steady levels of the metabolite despite flux changes.

5.2.4. Proportional Activation Analysis of Glycogen Synthesis

In order to solve Equation (5.5) it is necessary to possess values for the change in pathway flux and metabolite concentration, and for the elasticities of subsystems A and B with respect to M under varying conditions of the external parameter, X. Given our interest in the role of insulin signaling in maintaining homeostasis, we have selected insulin as the external (X) parameter. Under conditions of elevated insulin (infused at 10 mU/kg min) the elasticities of GT/HK (ε_{G6P}^A) and GSase (ε_{G6P}^B) with respect to G6P have been

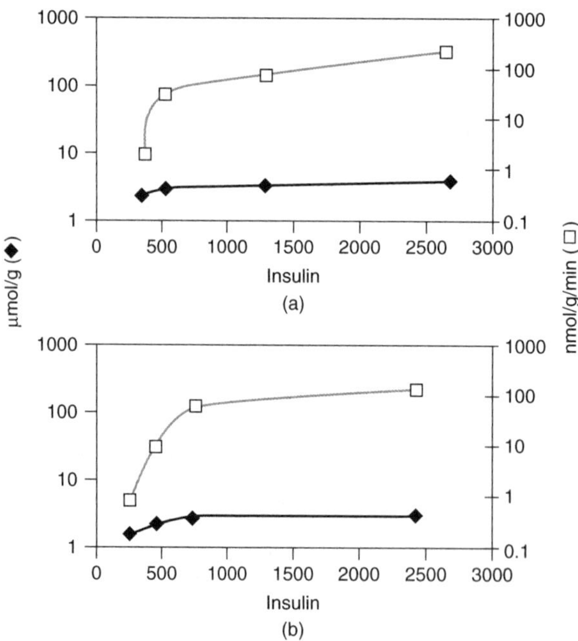

Figure 5.2. Insulin effects on glycogen synthase and [G6P]. Figure adapted from Rossetti and Hu (24) describing the changes in [G6P] (♦) and glycogen synthetic rate (□) at constant plasma glucose while plasma insulin was elevated from (a) 350 to 2700, or (b) 250 to 2400 pmol in rats that had been fasted for (a) 6 h or (b) 24 h. While [G6P] increases less than (a) 1.7-fold or (b) 3-fold, glycogen synthesis increases more than (a) 30-fold or (b) 50-fold. Note the different scales.

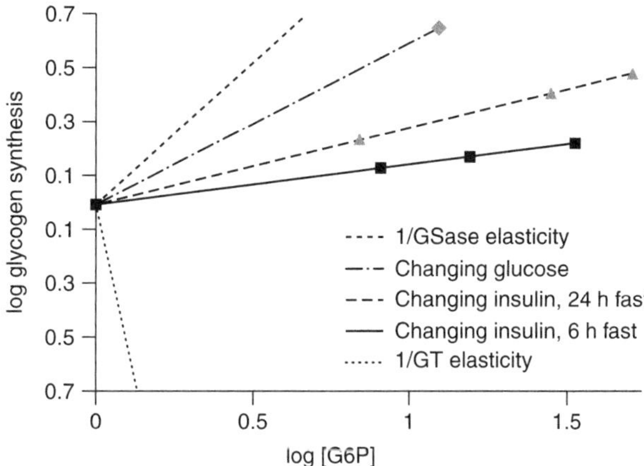

Figure 5.3. Glycogen synthase and glucose transport co-responsivity. The co-response coefficients (τ values) for the conditions of constant insulin with increasing glucose [values from Chase *et al.* (21)] and constant glucose with increasing insulin [values from Rossetti and Hu (24)] are given by the slopes of the lines on this plot. The dotted lines plot the extreme τ values of $1/\varepsilon^{A}_{\mathrm{G6P}}$ and $1/\varepsilon^{Binvitro}_{\mathrm{G6P}}$. The observed τ value for changing glucose (21), equaling $1/\varepsilon^{Binvivo}_{\mathrm{G6P}}$ (see Methods), is the alternative reference value for calculating the π^{AB}_{X} values for insulin. Its closeness to the extreme condition of $1/\varepsilon^{Binvitro}$ is additional support for the hypothesis that glucose affects flux primarily via GT/HK and thus results in poor G6P homeostasis. The τ values for changing insulin are much closer to a null value, indicating distributed control and comparatively good G6P homeostasis. The flux and concentration values from the different experiments have been normalized relative to unit flux and G6P concentrations in the controls.

derived previously (21). Chase *et al.* (21) determined the glucose transport/hexokinase (GT/HK) elasticity for a range of glycolytic fluxes. Since Equation (5.4) is based on a linear pathway, and since glycolytic shunting from the glycogen synthetic pathway is low during high fluxes, we used the elasticity value of −0.185, determined for the lowest glycolytic flux. This value is consistent with NMR measurements of this flux by Jucker *et al.* (19). The *in vitro* elasticity of GSase was reported as 0.79; however this was a misprint (D.L. Rothman, personal communication) and we are using the corrected value of 0.97. These result in reciprocal elasticities (that is, potential limiting values of τ for $\pi^{AB}_{\mathrm{G6P}} = \infty$ and 0 respectively) of −5.41 for the GT/HK subsystem and 1.03 for the GSase subsystem. These latter values appear as the reference limiting slopes shown in Figure 5.3. The *in vivo* elasticity of GSase under the conditions of low glycolytic flux was reported to be 1.9, resulting in a reciprocal elasticity of 0.53. We performed the proportionality analysis using both the *in vitro* and *in vivo* elasticities, and the impact of the differences on the calculated results is discussed later in the text.

The value of τ was calculated from the slope of the plots of log $V_{\mathrm{glycogen\ synthesis}}$ vs log [G6P] (Figure 5.3), using data obtained by Rossetti and Hu (24), and Chase *et al.* (21) over a range of insulin levels with a constant level of glucose [Figure 5.2(a,b)]. In both cases the animals had been fasted prior to the experiment to simulate the conditions of the transition from fasting to fed states.

5.3. RESULTS

5.3.1. *X* = Insulin: Calculation of Proportional Activation Using *In Vitro* and *In Vivo* Elasticities

In the experiments by Rossetti and Hu (24), fasted animals were exposed to stepwise increases in plasma insulin with constant plasma glucose. Both [G6P] and the glycogen synthesis rate rose with higher insulin [see Figure 5.2(a, b)]. However, while the [G6P] rise over the full range of insulin levels ranged from ~1.6- to 3-fold, the increase in glycogen synthesis rate was a much greater, ~35- to 50-fold. Since the [G6P] changes were very small relative to those in glycogen synthesis, the τ values were small (0.15 and 0.28 for the 6 and 24 h fasted rats, respectively) and the π values, 0.83 and 0.69 (using *in vitro* elasticities) and 0.7 and 0.44 (using *in vivo* elasticities), were in the range where it must be concluded there is a significant activation of the glycogen synthase subsystem in addition to that of the GT/HK subsystem (Figure 5.3). That is, insulin acts on the subsystems both up- and downstream of G6P – GT/HK and GSase – directly and to moderately comparable degrees, rather than exclusively through feed-back or feed-forward systems that make use of changes in G6P concentration. Qualitatively, this means that insulin is able to effect flux changes with minimal perturbation of G6P homeostasis.

5.3.2. Effect of Phosphorylation State on the Elasticity of Glycogen Synthase

A limitation of using the Chase *et al.* (21) elasticity results are that they were measured for one value of insulin, while the data used to calculate τ was obtained over the full range of physiological through supraphysiological insulin concentrations. Examination of Figure 5.3 shows that the value of τ is low (and, within accuracy, constant) throughout the range examined, which means that, provided the elasticities of the system do not change greatly with insulin, the conclusion of a proportional activation near unity is valid throughout the range of insulin stimulated GSase activation. In order to assess whether there could be large changes in the elasticity of GSase, we analyzed the *in vitro* results of GSase velocity vs G6P as a function of the degree of GSase phosphorylation as plotted by Roach and Larner (28). These measurements were performed under similar conditions, designed to mimic the *in vivo* milieu, to the *in vitro* measurements of the Chase *et al.* (21) study. The elasticities calculated from the $\%I$ range induced by physiological levels of insulin levels (less than ~40 $\%I$) were 1.2 (~10 $\%I$) and 0.84 (~20 $\%I$) – both comparable to the Chase *et al.* (21) value of 0.97 measured at ~40 $\%I$. At ~60 $\%I$ GSase elasticity dropped to 0.20, suggesting that at supraphysiologic levels of insulin GSase may be even less sensitive to G6P than our model suggests and that our prediction is a minimum estimate of the amount of proportional activation seen between GSase and GT. Homeostasis may depend even less on feed-forward activation of GSase than our conservative interpretation indicates.

5.4. DISCUSSION

Our results, in conjunction with the previous control analysis presented by Chase *et al.* (21) have implications specifically for the glycogen synthesis pathway and for metabolic control in general. The conditions of high glucose and variable insulin that we examined grossly reflect the two-step transition from the fasting to the fed state in mammals. In the fasting state, both blood glucose and plasma insulin were low – reflecting the decreased environmental availability of glucose and the attendant lack of need for insulin control. With feeding there was a rapid rise in plasma glucose that initially outpaced the ability of the pancreas to release insulin. Several minutes after the onset of feeding the pancreas began to release insulin in substantial quantities into the bloodstream, resulting in a fall of plasma glucose concentration to close to pre-meal values. For example, in the case of healthy human subjects given a large mixed meal, plasma glucose rose from 5.4 to 7.3 mmol in the first 30 min after a meal followed by a rapid fall to 6.4 mmol at 75 min and a progressive decrease to pre-meal levels at 240 min (29, 30). During the high insulin phase there was a large increase in muscle glycogen synthesis (29, 30), as observed in studies under controlled high-insulin conditions such as the Chase *et al.* (21) and Rosetti and Hu (24) papers used in the present analysis. The increase in glycogen synthesis removes excess glucose from the blood and plays a key role in maintaining plasma glucose homeostasis. In type 2 diabetes, where insulin-stimulated muscle glycogen synthesis was reduced as much as two-fold (30, 31) plasma glucose levels were as high as 10 mM, even 240 min after a meal (30). Despite the many-fold increase in glycogen synthesis there were only small insulin-stimulated changes in G6P (e.g. 11, 18, 19, 21, 24, 32, 33). Looking at these dual requirements of the muscle glycogen synthesis pathway – to maintain plasma glucose and cellular G6P homeostasis together – we propose below a model for how the kinetic properties of the enzymes of the muscle glycogen synthesis system, and the alterations in their activity with insulin signaling, meet these simultaneous demands

5.4.1. Role of Insulin-stimulated Proportional Activation in Maintaining Plasma Glucose and Muscle G6P Homeostasis during the Steady-state Absorption Period

Release of insulin by the pancreas during glucose absorption leads to the coordinate activation of glucose transport, hexokinase and glycogen synthase via insulin signaling pathways. However, given that glucose transport has the dominant flux control coefficient and the elasticity of GSase is high even at low phosphorylation levels, it is not necessary to activate GSase in order to have the glycogen synthesis rate increase sufficiently to match glucose absorption into the blood. The importance of the coordinate activation in maintaining both plasma glucose and G6P homeostasis may be appreciated by examining the consequences of no activation of GSase.

At a given insulin level, glucose concentration can only increase muscle glycogen synthesis linearly, due to the kinetic properties of glucose transport and hexokinase (16). Because the rate of glucose influx from the digestive system can be high and last for several hours, plasma glucose concentrations would not activate muscle glycogen synthesis sufficiently to reach a steady state until glucose rose to many times fasting levels. G6P levels would rise proportionately to glucose, leading to severe alterations in cellular homeostasis and possibly transport control due to feedback inhibition on hexokinase. This phenomenon is evidenced by the extraordinarily high plasma glucose concentrations achieved after a meal by insulin-dependent diabetes mellitus patients, who cannot produce insulin.

Insulin stimulation of the GT/HK and GSase portions of the muscle glycogen synthesis pathway, therefore, is necessary to deal with substantial sustained glucose influx while maintaining plasma glucose and cellular G6P homeostasis. As insulin concentration rises, glycogen synthesis may increase more than 30-fold, an order of magnitude different from when synthesis is stimulated by glucose alone. However, the concentration of G6P rises less than two-fold across this range, and plasma glucose concentration is maintained at only 20 % above pre-meal levels. From the axiom of Kacser and Acarenza (13), for metabolite concentration to be maintained constant despite such a precipitous rise in flux the enzymes, both upstream *and* downstream of the metabolite must increase their activity in parallel.

It should be noted that we did not observe a value of π^{AB}_{G6P} of 1.0, but a slightly lower value, suggesting somewhat imperfect proportional activation. For a value of π^{AB}_{G6P} of 0.7, it can be seen from Equation (5.6) that, the larger the value of the elasticity of GSase to G6P, $\varepsilon^{B}_{\mathrm{G6P}}$, the closer τ, the co-response coefficient of M and J to X, will approach zero. In other words, the cooperative response of GSase to its effector G6P is still making some contribution to the homeostasis of G6P levels, whereas, had π^{AB}_{G6P} been unity, the value of this elasticity would be irrelevant.

5.4.2. The Roles of Insulin-induced Phosphorylation and Allostery in the Proportional Activation of GSase

Insulin has been shown to increase GT activity by moving storage vesicles to the plasma membrane. However, were this the only way for it to increase glycogen synthesis, then G6P concentration would have to be considerably higher and the π value would approach zero. Since the π value approaches unity, insulin signaling must also directly stimulate an increase in GSase activity to match the increase in GT flux. It is well established that insulin, through signaling pathways not yet fully defined, also leads to a rapid (minutes) change in the phosphorylation state of GSase, which alters its kinetics (34). Based on the *in vitro* studies of Chase *et al.* (21), as well as earlier studies of Roach and Larner (28), the change in phosphorylation state is sufficient to explain most of the increase in activity with insulin stimulation, and, as the present analysis shows, leads to a π value close to unity. However, the *in vivo* elasticity measured by Chase *et al.* (21) is higher than the *in vitro* value, suggesting the possibility of additional mechanisms for activating GSase. These may include a dual role of G6P as a substrate of the block and as an allosteric effector of GSase. Although the latter term is likely to be dominant since elasticities of cooperative effects are larger than simple substrate elasticities, the response of GSase may have a component due to an increase of UDP-glucose, which is hidden within the GSase block of enzymes and metabolites considered in the control analysis, but which will be correlated with the G6P level. Since the *in vivo* levels are well below the GSase K_m, if the concentration of UDP-glucose increased proportionately with [G6P] it would explain the higher *in vivo* elasticity without the need to postulate an additional mechanism of GSase activation.

5.4.3. Relevance to Human Insulin-stimulated Glucose Metabolism and Diabetes

While this study evaluated data from rodents, we have some reason to believe that it may be applicable to human subjects at well. Specifically, even under experimentally induced conditions of hyperglycemia and hyperinsulinemia, healthy human subjects never displayed a greater than two-fold increase in [G6P] (11, 32, 33, 35, 36) – suggesting the importance of G6P homeostasis in human skeletal muscle. Given the delicacy of the system's homeostatic balance, it is surprising that it does not appear to be substantially disrupted in patients with NIDDM or genetic obesity and lipid-induced insulin resistance (11, 32, 33, 36). This finding has implications for our understanding of diabetes pathology and pharmacology. Specifically, since NIDDM and insulin resistant patients are able to maintain G6P homeostasis, both subsystems of the glycogen synthetic pathway must be adequately functioning and responsive (although down-regulated). Thus, defects in glycogen synthesis appear to be the result of balanced, but insufficient, signaling by insulin rather than from localized defects in particular pathway enzymes. This conclusion is consistent with the finding of impaired signaling protein activity in both insulin-resistant syndromes and lipid-induced insulin resistance [for a review see Peterson and Shulman (37)]. This suggests that investigations of diabetes pathology and pharmacologic interventions should be focused on upstream insulin signaling. Furthermore, it suggests that pharmacological approaches that modulate isolated enzymes may not only fail to address the underlying pathophysiology, but may also worsen conditions by unlinking GSase and GT/HK and disrupting previously good homeostasis. However, it should be noted that, while healthy individuals maintained good homeostasis, a small fraction of insulin resistant subjects lost homeostasis, a result that was not observed when averaged with their peers (D.L. Rothman, personal communication). Thus, while interventions aimed at particular enzymes may not be effective for addressing most patients, certain individual can be selected who are likely to benefit from such treatments.

5.4.4. Phosphorylation as a General Homeostatic Mechanism

Presently, changes in gene expression during changes of state are intensely studied in organisms ranging from bacteria to humans. Many of the gene products have been shown to be kinases and to change degrees of phosphorylation of their targets. In the absence of another function for enzymatic phosphorylation, it is assumed that flux is being controlled. The present results, showing that phosphorylation of GSase serves to maintain homeostasis, not to control flux, challenge the universality of that assumption and offer an alternate function. To the extent that biochemical mechanisms are postulated upon the assumption that kinase activity controls flux, caution must be exercised in view of the present results. An equivalent point has been made by Hofmeyr and Cornish-Bowden (4, 5) with respect to the greater importance of the cooperativity of allosteric enzymes in metabolite homeostasis than in flux control.

Given the success of the glycogen synthetic pathway at coordinating internal and external homeostasis, it seems likely that there are analogous arrangements in other pathways – at least in storage pathways that concentrate flux control on the supply side in order to coordinate storage with nutrient availability. Specifically, we predict the following as a general mechanism for the linkage of internal and external homeostasis at a variety of time scales, particularly for supply-controlled pathways. Rapid and small-scale flux changes will be controlled by a sub-system that acts by mass action and thus responds primarily to changes in the external milieu. The activity of its paired sub-systems will be coordinated by allosteric modulation – at the expense of perfectly maintained internal homeostasis. Larger-scale flux changes operating on a more intermediate time scale will involve an external detector/effector (e.g. pancreas/insulin) that stimulates both up- and downstream sub-systems, thereby maintaining excellent internal *and* external homeostasis despite increased flux. However, as it relies on an external detector, it is necessarily slower and requires a more complex mechanism to stimulate the two sub-systems equivalently. In the example discussed above, this stimulation is achieved through phosphorylation. Other mechanisms could realize the same end but there is every reason to expect that phosphorylation will serve other pathways similarly, although for operations of a much longer time scale there remains the well-described method of altering gene expression. While this is an excellent way to pair multiple sub-systems, it is necessarily the slowest and most coarse homeostatic and flux modulating mechanism.

As dynamic *in vivo* measurements become increasingly available from multicellular organisms that must manage homeostasis in different compartments, we anticipate an increased interest in understanding how pathways are able to meet the dual requirements of changing flux while maintaining homeostasis. The complex phosphorylation states and allosteric sensitivity of GSase, engineered to serve metabolite homeostasis rather than flux control, provides one well-understood example of the realization of those goals.

Acknowledgments

This work was supported by NIH grant DK27121 and MSTP training grant GM07205, and will be submitted in part to fulfill the requirements for the Degree of Doctor of Philosophy at Yale University.

REFERENCES

1. Kacser, H. and Burns, J.A. (1973) The control of flux. *Symp. Soc. Exp. Biol.* **27**, 65–104. [Reprinted as: Kacser, H., Burns, J.A. and Fell, D.A. (1985) *Biochem. Soc. Trans.* **23**, 341–366.]
2. Heinrich, R. and Rapoport, T.A. (1974) A linear steady-state treatment of enzymatic chains. General properties, control and effector strength. *Eur. J. Biochem.* **42**, 89–95.
3. Fell, D.A. (1997) *Understanding the Control of Metabolism* (Portland Press, London).
4. Hofmeyr, J.H. and Cornish-Bowden, A. (1991) Quantitative assessment of regulation in metabolic systems. *Eur. J. Biochem.* **200**, 223–236.
5. Hofmeyr, J.H. and Cornish-Bowden, A. (2000) Regulating the cellular economy of supply and demand. *FEBS Lett.* **476**, 47–51.
6. Fell, D.A. and Thomas, S. (1995) Physiological control of metabolic flux: the requirement for multisite modulation. *Biochem. J.* **311**, 35–39.
7. Korzeniewski, B., Harper, M.E. and Brand, M.D. (1995) Proportional activation coefficients during stimulation of oxidative phosphorylation by lactate and pyruvate or by vasopressin. *Biochim. Biophys. Acta* **1229**, 315–322.
8. Thomas, S. and Fell, D.A. (1996) Design of metabolic control for large flux changes. *J. Theor. Biol.* **182**, 285–298.
9. Hochachka, P.W. (1994) *Muscles as Molecular and Metabolic Machines* (CRC Press, Boca Raton, FL).

10. Thomas, S. and Fell, D.A. (1998) The role of multiple enzyme activation in metabolic flux control. *Adv. Enzyme Regul.* **38**, 65–85.
11. Rothman, D.L., Shulman, R.G. and Shulman G.I. (1992) ^{31}P nuclear magnetic resonance measurements of muscle glucose-6-phosphate. Evidence for reduced insulin-dependent muscle glucose transport or phosphorylation activity in non-insulin-dependent diabetes mellitus. *J. Clin. Invest.* **89**, 1069–1072.
12. Shulman R.G. and Rothman D.L. (2001) ^{13}C NMR of intermediary metabolism: implications for systemic physiology. *A. Rev. Phys.* **63**, 15–48.
13. Kacser, H. and Acerenza, L. (1993) A universal method for achieving increases in metabolite production. *Eur. J. Biochem.* **216**, 361–367.
14. Fell, D.A. (2000) Signal transduction and the control of expression of enzyme activity. *Adv. Enzyme Regul.* **40**, 35–46.
15. Shulman R.G. and Rothman D.L. (1996) Enzymatic phosphorylation of muscle glycogen synthase: a mechanism for maintenance of metabolic homeostasis. *Proc. Natl Acad. Sci. USA* **93**, 7491–7495.
16. Shulman, R.G., Bloch, G. and Rothman, D.L. (1995) *In vivo* regulation of muscle glycogen synthase and the control of glycogen synthesis. *Proc. Natl Acad. Sci. USA* **92**, 8535–8542.
17. Roach, P.J. (2002) Glycogen and its metabolism. *Curr. Mol. Med.* **2**, 101–120.
18. Jucker, B.M., Rennings, A.J., Cline, G.W. and Shulman, G.I. (1997) ^{13}C and ^{31}P NMR studies on the effects of increased plasma free fatty acids on intramuscular glucose metabolism in the awake rat. *J. Biol. Chem.* **272**, 10464–10473.
19. Jucker, B.M., Rennings, A.J., Cline, G.W., Peterson, K.F. and Shulman, G.I. (1997) *In vivo* NMR investigation of intramuscular glucose metabolism in conscious rats. *Am. J. Physiol.* **272**, E139–148.
20. Cline, G.W., Petersen, K.F., Krssak, M., Shen, J., Hundal, R.S., Trajanoski, Z., Inzucchi, S., Dresner, A., Rothman, D.L. and Shulman, G.I. (1999) Impaired glucose transport as a cause of decreased insulin-stimulated muscle glycogen synthesis in type 2 diabetes. *New Engl. J. Med.* **341**, 240–246.
21. Chase, J.R., Rothman, D.L. and Shulman, R.G. (2001) Flux control in the rat gastrocnemius glycogen synthesis pathway by *in vivo* ^{13}C/^{31}P NMR spectroscopy. *Am. J. Physiol. Endocrinol. Metab.* **280**, E598–E607.
22. Cline, G.W., Johnson, K., Regittnig, W., Perret, P., Tozzo, E., Xiao, L., Damico, C. and Shulman, G.I. (2002) Effects of a novel glycogen synthase kinase-3 inhibitor on insulin-stimulated glucose metabolism in Zucker diabetic fatty (*fa*/*fa*) rats. *Diabetes* **51**, 2903–2910.
23. Agius, L., Peak, M., Newgard, C.B., Gomez-Foix, A.M. and Guinovart, J.J. (1996) Evidence for a role of glucose-induced translocation of glucokinase in the control of hepatic glycogen synthesis. *J. Biol. Chem.* **271**, 30479–30486.
24. Rossetti, L. and Hu, M. (1993) Skeletal muscle glycogenolysis is more sensitive to insulin than is glucose transport/phosphorylation. Relation to the insulin-mediated inhibition of hepatic glucose production. *J. Clin. Invest.* **92**, 2963–2974.
25. Fell, D.A. and Sauro, H.M. (1985) Metabolic control and its analysis. Additional relationships between elasticities and control coefficients. *Eur. J. Biochem.* **148**, 555–561.
26. Brown, G.C., Hafner, R.P. and Brand, M.D. (1990) A 'top-down' approach to the determination of control coefficients in metabolic control theory. *Eur. J. Biochem.* **188**, 321–325.
27. Hofmeyr, J.H. and Cornish-Bowden, A. (1996) Co-response analysis: a new experimental strategy for metabolic control analysis. *J. Theor. Biol.* **182**, 371–380.
28. Roach, P.J. and Larner, J. (1976) Rabbit skeletal muscle glycogen synthase. II. Enzyme phosphorylation state and effector concentrations as interacting control parameters. *J. Biol. Chem.* **251**, 1920–1926.
29. Taylor, R., Price, T.B., Katz, L.D., Shulman, R.G. and Shulman, G.I. (1993) Direct measurement of change in muscle glycogen concentration after a mixed meal in normal subjects. *Am. J. Physiol.* **265**, E224–E229.
30. Carey, P.E., Halliday, J., Snaar, J.E.M., Morris, P.G. and Taylor, R. (2003) Direct assessment of muscle glycogen storage after mixed meals in normal and type 2 diabetic subjects. *Am. J. Physiol.* **284**, E688–E694.
31. Shulman, G.I., Rothman, D.L., Jue, T., Stein, P., DeFronzo, R.A. and Shulman, R.G. (1990) Quantitation of muscle glycogen synthesis in normal subjects and subjects with non-insulin-dependent diabetes by ^{13}C nuclear magnetic resonance spectroscopy. *New Engl. J. Med.* **322**, 223–228.
32. Rothman, D.L., Magnusson, I., Cline, G., Gerard, D., Kahn, C.R., Shulman, R.G. and Shulman, G.I. (1995) Decreased muscle glucose transport/phosphorylation is an early defect in the pathogenesis of non-insulin-dependent diabetes mellitus. *Proc. Natl Acad. Sci. USA* **92**, 983–987.

33. Perseghin, G., Price, T.B., Petersen, K.F., Roden, M., Cline, G.W., Gerow, K., Rothman, D.L. and Shulman, G.I. (1996). Increased glucose transport-phosphorylation and muscle glycogen synthesis after exercise training in insulin-resistant subjects. *New Engl. J. Med.* **335**, 1357–1362.
34. Lawrence, J.C. and Roach, P.J. (1997) New insights into the role and mechanism of glycogen synthase activation by insulin. *Diabetes* **46**, 541–547.
35. Roden, M., Price, T.B., Perseghin, G., Petersen, K.F., Rothman, D.L., Cline, G.W. and Shulman, G.I. (1996) Mechanism of free fatty acid-induced insulin resistance in humans. *J. Clin. Invest.* **97**, 2859–2865.
36. Petersen, K.F., Hendler, R., Price, T., Perseghin, G., Rothman, D.L., Held, N., Amatruda, J.M. and Shulman, G.I. (1998) $^{13}C/^{31}P$ NMR studies on the mechanism of insulin resistance in obesity. *Diabetes* **47**, 381–386.
37. Petersen, K.F. and Shulman, G.I. (2002). Pathogenesis of skeletal muscle insulin resistance in type 2 diabetes mellitus. *Am. J. Cardiol.* **90**, 11G–18G.

6

Regulation of Glycogen Metabolism in Muscle during Exercise

Thomas B. Price

Yale University School of Medicine, Department of Diagnostic Radiology, PO Box 208042, New Haven, CT 06520-8042, USA

6.1. INTRODUCTION

The human musculature is well suited to NMR for several reasons. First, insomuch as the musculature is the primary organ of carbohydrate uptake and storage (Taylor *et al.*, 1993), it depends upon glucose and glycogen, both of which give strong ^{13}C NMR signals. The musculature is also the organ of physical

Metabolomics by In Vivo NMR. Edited by R. G. Shulman and D. L. Rothman
 ISBN: 0-470-84719-0

work, and as such is the primary organ of carbohydrate utilization. Because glycogen and glucose can be measured with ^{13}C NMR, both carbohydrate storage and carbohydrate utilization can be studied.

Second, using localized NMR an individual muscle can be non-invasively studied in the body while systemic metabolic and hormonal conditions can be controlled. This ability to assess 'local' metabolic events can simplify data analysis considerably.

Finally, exercise is a natural metabolic stimulus that consumes muscle glycogen. Because both exercise and NMR are non-invasive, many measurements can be taken over a short period of time and accurate rates of glycogen depletion and recovery can be calculated. A muscle's ability to restore its glycogen reserves following exercise is strongly influenced by numerous factors such as obesity, lifestyle and disease. Thus, by measuring these rates NMR can provide insight into the muscle's biochemical health.

In addition to ^{13}C muscle measurements, creatine phosphate (PCr), inorganic phosphate (Pi), and adenosine triphosphate (ATP) are all measurable with ^{31}P techniques (Gadian *et al.*, 1976; Dawson *et al.*, 1980; Chance *et al.*, 1981). ^{31}P NMR also provides data from which muscle pH can be calculated (Ogawa *et al.*, 1978). Glucose-6-phosphate (G6P) is an intermediary metabolite that is of interest to muscle carbohydrate balance, and with effort it can also be measured with ^{31}P NMR. By using a combination of ^{13}C NMR to measure glycogen storage and utilization and ^{31}P NMR to measure G6P and the high-energy phosphates, the measurable range of metabolic events can be expanded.

This chapter focuses on NMR studies of muscle carbohydrate metabolism, and as such will discuss primarily glycogen measurements. We begin by examining the nature of muscle glycogen and its assessment by NMR and biochemical methods. We then discuss NMR studies of intense local exercise/recovery. The chapter then focuses on MRI as a tool for making the transition from non-invasive carbohydrate study of local exercise to non-invasive study of systemic carbohydrate metabolism. Current studies continue to bridge the gap between local exercise and exercise of groups and systems of muscles.

6.2. MUSCLE BASICS

6.2.1. NMR and Muscle Glycogen

In order for NMR to become a viable tool for studying muscle glycogen it was first necessary to demonstrate that NMR could 'see' glycogen. Early ^{13}C NMR measurements had identified peaks at 100.5 ppm as the C_1 of glycogen (Griffin *et al.*, 1975; Cohen *et al.*, 1979). The visibility of glycogen was established by Sillerud and Shulman (1983), who reported that ^{13}C NMR spectra of glycogen in solution gave sharp resonances from $\sim$100 % of the carbon nuclei. A subsequent study demonstrated that these resonances were also visible in the perfused rat liver (Shulman *et al.*, 1988). Proof of the visibility has been necessary because a macromolecule the size of glycogen would not give sharp visible lines if it were rigid. In subsequent NMR studies of relaxation times it was demonstrated that the large glycogen molecules had extensive internal motion (Zang *et al.*, 1990a, b), probably owing to its highly branched and multitiered organization. A more recent study demonstrated that muscle glycogen was 100 % visible in a skinned rabbit muscle preparation (Gruetter *et al.*, 1991). In a human population, an NMR validation study focused on repetitive 5 min measurements of the same muscle to evaluate reproducibility (Taylor *et al.*, 1992). The study reported a coefficient of variation that was $\pm$4.3 %, which for the basal muscle glycogen level of $\sim$80 mM was $\pm$3.4 mM. As expected, the accuracy of these measurements depended upon the NMR signal-to-noise ratio, which improved as the square root of the time of measurement. Several measurements with 20 min acquisition yielded a 2$\times$ smaller coefficient of variation than expected (Price *et al.*, 1991). However, the validation study showed that, even with 5 min data acquisitions, the accuracy was significantly better than that reported from biopsy measurements (Taylor *et al.*, 1992). This improved time resolution was pivotal to the study of muscle glycogen recovery following severe glycogen depletion. During these experiments it became clear that the errors were due to the signal-to-noise of the NMR spectra and no additional uncertainties were introduced when the subjects left the magnet between measurements. Since optimum data acquisition conditions of magnet shimming and probe tuning could be reached within several minutes, there were no serious disadvantages in allowing the subject to leave the magnet between data acquisitions. As suitable NMR equipment has become available, well-resolved ^{13}C resonances have been regularly observed *in vivo* in both animal and human muscle and liver (Shulman *et al.*, 1990; Price *et al.*, 1994a; Taylor *et al.*, 1996).

A simple ^{1}H de-coupled pulse/acquire pulse program is used to obtain ^{13}C NMR spectra of skeletal muscles (Price *et al.*, 1991; Taylor *et al.*, 1992). A surface coil radiofrequency (RF) probe is positioned at the skin surface adjacent to the muscle of interest. All information about glycogen is obtained from the spectral resonance at 100.5 ppm that corresponds to the 1-^{13}C glucose atoms in the glycogen-bound form (Taylor *et al.*, 1992). The integral area beneath this resonance is a direct indication of the number of 1-^{13}C glycogen-bound glucose atoms present. Conversely, the 1-^{13}C free-glucose atom resonates at two different chemical shift frequencies, one corresponding to the α-glucose anomer (92 ppm) and the other to β-glucose (96 ppm). Chemical concentrations of glycogen are determined by comparing each human *in vivo* spectrum with a spectrum that is obtained from an external standard solution of known glycogen concentration (150–200 mM). As discussed in Chapter 2, NMR-determined muscle glycogen concentrations have been found to agree well with concentrations determined by direct biochemical assay of muscle samples obtained from the same subject and on the same day (Taylor *et al.*, 1992).

NMR studies of exercise-induced changes in muscle glycogen would be invalid if the exercise produced a transient non-concentration-related loss of the NMR signal. Several exercise-related factors that could potentially affect the NMR signal have been studied. These include: (1) intramuscular temperature; (2) ionic strength; and (3) water content. Measurements of glycogen in solution have shown no significant difference in the 1-^{13}C glycogen T_1 between 295 and 310 K (Zang *et al.*, 1990b). In addition, any exercise-induced temperature increase would be expected to dissipate rapidly (on the order of a couple of minutes) as a result of increases in subcutaneous blood flow (Nielsen *et al.*, 1988), while glycogen recovers over an initial period of about 60 min (Price *et al.*, 1994a, 1996, 2003a). Changes in coil loading due to exercise-induced changes in muscle ionic concentration have been monitored and found to be insignificant (Price *et al.*, 1994a). Muscle volume swelling immediately after plantar flexion exercise has also been monitored by measuring the increase in calf circumference and found to be <4 %, decreasing to <1 % within 5 min (Price *et al.*, 1994a).

6.2.2. Biochemical Analysis of Muscle Glycogen

Previously, information about the complicated patterns of muscle glycogen balance and its relation to exercise was only available from needle biopsies that could directly measure muscle glycogen concentrations (Dietrichson *et al.*, 1987; Harris *et al.*, 1974). Numerous biopsy studies provided the original broad outline of muscle glycogen metabolism. While the traditional biopsy method of muscle glycogen assay can cause discomfort for the subject, thereby limiting the number of measurements and increasing the time between measurements, information is available from biopsy samples that NMR cannot provide. Information about metabolic events such as enzyme activities and rates, concentrations of various proteins, and mRNA synthesis are invaluable to understanding the complicated metabolic processes associated with muscle exercise and recovery. Because muscle biopsies sample only a small volume (50–80 mg) in a non-homogeneous tissue and glycogen analysis requires a sizable portion of each biopsy sample (40–50 mg), not much tissue is available for other analyses. Previous studies, recognizing the difficulties of glycogen determination by biopsy, had measured glucose uptake (Fell *et al.*, 1982). The conversion of glucose uptake to glycogen synthesis depended upon assumptions about flux division that introduced possible errors into the conclusions. In the experiments where glycogen was measured by biopsy, the number of biopsies seriously circumscribed any conclusions that could be drawn regarding glycogen synthesis/depletion rates (Maehlum *et al.*, 1977; Maehlum and Hermansen, 1978).

6.3. NMR TO STUDY LOCAL MUSCLE METABOLISM

6.3.1. The Appeal of Local Exercise

^{13}C NMR was first applied in human exercise studies to measure glycogen levels in the gastrocnemius muscles of two athletic males before and after they ran a half-marathon (Avison *et al.*, 1988). The exercise resulted in a 70 % decrease in muscle glycogen concentration that returned to 80 % of baseline during the next 19 h. Although this first exercise/NMR experiment employed a systemic exercise protocol (Avison *et al.*, 1988), most of the subsequent early studies utilized localized exercise (Price *et al.*, 1991, 1994a, b).

The type of exercise protocol is of primary importance in determining the source of substrates during exercise. McDermott *et al.* (1991) have shown in rats that, when a large portion of the total musculature is exercised, glycogen levels decrease in non-exercising muscles. This study reported up to a 37 % decrease in suspended hind limb muscle glycogen during 90 min of forelimb treadmill exercise by male Sprague–Dawley rats (McDermott *et al.*, 1991). Exercise of large muscle groups induces a number of metabolic and hormonal responses that can affect an individual muscle's metabolism during exercise. Another effect of systemic exercise is that, during prolonged exercise of groups of muscles the contribution of an individual muscle to the total workload may change, significantly complicating any attempt at analysis of local metabolism.

The appeal of using a localized exercise protocol starting from a basal steady-state condition to study muscle glycogen metabolism by NMR is that the systemic response to localized exercise is buffered. Early NMR studies suggested that, during localized exercise, glycogen synthesis in the exercised muscle is under local control. In the initial study glycogen was measured in the right and left gastrocnemius muscles before and after a single leg–toe raise protocol that involved only the right leg (Price *et al.*, 1991). While the muscle in the exercised right leg had substantially reduced glycogen levels, the contralateral muscle was not affected by the exercise. Following exercise, glycogen concentrations increased over several hours towards a basal level in the right muscle while the unexercised muscle, perfused by the same blood, did not synthesize any glycogen. A subsequent study infused 99 % labeled 1-^{13}C glucose to improve the ^{13}C NMR signal, demonstrating that there was no turnover in the unexercised leg (Price *et al.*, 1994b). The 1-^{13}C labeling technique, by increasing the 1-^{13}C signal, enables small (tenths of mM) glycogen fluxes to be observed that would not be detectable by natural abundance ^{13}C NMR.

6.3.2. Assessing the Workload

The classic criterion for assessing human exercise workloads is percentage maximum oxygen uptake volume ($\%VO_{2\,max}$) (Hermansen *et al.*, 1967; Gollnick *et al.*, 1974; Vollestad and Blom, 1985; Ahlborg *et al.*, 1986). Workloads $<45\,\%VO_{2\,max}$ are considered light, while $45–80\,\%VO_{2\,max}$ is considered moderate and $>80\,\%VO_{2\,max}$ is heavy (Hermansen *et al.*, 1967; Gollnick *et al.*, 1974; Vollestad and Blom, 1985; Ahlborg *et al.*, 1986). Because the vast majority of exercise studies have employed the $\%VO_{2\,max}$ criterion, a means of relating $\%VO_{2\,max}$ to percentage maximum voluntary contraction (%MVC) was necessary to interpret the results of local exercise studies. Measurements comparing MVC with $VO_{2\,max}$ in exercising quadriceps (Saltin and Karlsson, 1971; Gollnick *et al.*, 1974; Price *et al.*, 1991) suggest that the gastrocnemius performing plantar flexion at 15 %MVC is working at an equivalent workload to the quadriceps performing bicycling exercise at $30\,\%VO_{2\,max}$. At 51 %MVC the gastrocnemius is working at an equivalent workload to the quadriceps at 100–120 $\%VO_{2\,max}$. Put another way, there is a roughly 1:2 correspondence between %MVC and $\%VO_{2\,max}$ (Price *et al.*, 1994).

NMR methods consider the muscle as a whole and do not directly address the possibility that synthesis and degradation may be occurring in different muscle fibres simultaneously. Systemic exercise studies have found that at light workloads of <45 $\%VO_{2\,max}$ (corresponding to <22–23 %MVC), glycogen depletion in slow-twitch (ST) fibers precedes depletion in fast-twitch (FT) fibers (Gollnick *et al.*, 1974; Vollestad and Blom, 1985). Gollnick *et al.* demonstrated that, after 3 h of pedaling exercise at 31 $\%VO_{2\,max}$, almost all of the 50 % of muscle glycogen remaining in the human quadriceps femoris was in FT fibers. Ivy *et al.* (1987) found that, when the workload was increased to moderate levels (>45 $\%VO_{2\,max}$), all muscle fibers shared contraction. Thus, when analyzing data that has been obtained from the whole muscle, workload and its relationship to muscle fiber type activity can be considered, although this has not yet been done for NMR data.

6.4. NMR STUDIES OF INTENSE LOCAL EXERCISE AND RECOVERY

The amount of muscle glycogen depleted during exercise depends upon the intensity and duration of the exercise (Gollnick *et al.*, 1974; Robergs *et al.*, 1991). At high intensities, glycogen is the primary energy substrate and the duration of exercise can be limited by the glycogen supply (Robergs *et al.*, 1991). During intense exercise, muscle glycogen may be depleted to 25 % of baseline concentration in as little as 10 min of intermittent exercise (Price *et al.*, 1994a). Following intense exercise, the rate of glycogen recovery had been thought to depend upon the extent of the depletion (Maehlum *et al.*, 1977; Maehlum and Hermansen, 1978; Fell *et al.*, 1982; Ivy, 1991). Recently, however, we have shown that, following intense exercise, it is the amount of glycogen that remains, rather than the amount of glycogen that was depleted, that exerts control over the initial period of glycogen recovery (Price *et al.*, 2000).

6.4.1. Local Control of Muscle Glycogen Recovery in Healthy Subjects

Our early NMR studies of muscle glycogen recovery from extensive exercise reported that, during the early period of fasting recovery, there was rapid glycogen synthesis (27 ± 5 mmol/l h). This rapid recovery slowed to a steady rate of circa one-tenth the initial rate (2.9 ± 0.8 mmol/l h) within an hour, and remained steady over several hours of continued fasting recovery (Price *et al.*, 1991, 1994a). The rapid synthetic phase occurred only when the muscle's glycogen concentration was below 35 mmol/l (Figure 6.1). Although rapid glycogen synthesis (~25 mmol/l h) had been induced in humans by infusing glucose after systemic exercise of large muscle groups (Ivy, 1991), this high rate of synthesis had not been reported following exercise of an isolated muscle, nor had it been seen without infusing substrate.

In a 1977 study of recovery from severe exercise by non-diabetic (control) and insulin-dependant diabetic (IDDM) subjects, Maehlum *et al.* (1977) reported glycogen levels that were determined from biopsies obtained at 2, 4, 6, 9 and 12 h of recovery from exhaustive bicycle exercise. While Maehlum *et al.* did not report glycogen synthesis rates, the reported data are sufficient to calculate rates with their published time resolution. In the control population, the rate during the first 2 h of recovery was 7.4 mM/h and during the next 4 h it dropped to 5.0 mM/h. If the same time points are used to calculate glycogen synthesis rates from data reported in our 1994 study, the rates are 5.0 mM/h during the first 2 h of recovery from severe exercise and 2.5 mM/h during the next 4 h (Price *et al.*, 1994a). However, because we were able to obtain glycogen concentrations every 15 min, the initial glycogen synthesis rate during the first 30 min of recovery was 27 ± 5 mM/h. After 1 h of recovery the glycogen synthesis rate dropped significantly ($p < 0.002$) to 2.9 ± 0.8 mM/h and remained constant thereafter (Price *et al.*, 1994a). In IDDM subjects given insulin after exercise, Maehlum *et al.* (1977) observed that glycogen synthesis was similar to that of control subjects,

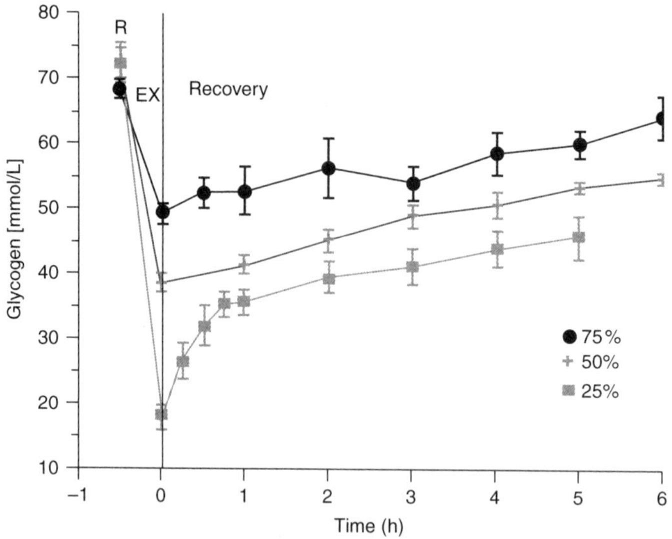

Figure 6.1. Fasting recovery of glycogen after depletion to 75, 50 and 25 % of baseline levels. Means ± SE. R = baseline at rest; EX = exercise. (Reproduced from Price, T.B., Rothman, D.L., Taylor, R., Shulman, G.I., Avison, M.J. and Shulman, R.G., *J. Appl. Physiol.* **76**(1): 104–111, 1994 by the American Physiological Society.)

but, when insulin was withheld, an initial phase of glycogen synthesis was observed which leveled off as concentrations reached 30–40 mM. With the improvement in time resolution available from ^{13}C NMR, we were able to document a biphasic synthesis in control subjects that had not been observed with muscle biopsy techniques.

Richter *et al.* in studies of rat muscle (Richter *et al.*, 1984; Garetto *et al.*, 1984) previously reported biphasic post-exercise recovery of muscle glycogen. In those studies, the initial rapid post-exercise rate of glycogen synthesis (soleus and gastrocnemius) declined ~6-fold within the first hour of recovery and remained constant thereafter, in agreement with our 1994 findings (Price, 1994). When the hindlimbs were placed in a perfused glucose medium with added insulin following exercise, a similar biphasic recovery was observed (Richter *et al.*, 1984). However, in the absence of insulin only the initial rapid phase was observed. Richter *et al.* (1984) concluded that the initial glycogen resynthesis following exercise was modulated by local contraction-induced factors. The local exercise protocol employed in our 1994 study produced minimal changes in blood metabolite and hormone levels, further supporting Richter's conclusions.

Animal studies have demonstrated increases in both the number and activity of Glut4 glucose transporters on the muscle plasma membrane following exercise (Douen *et al.*, 1990; Goodyear *et al.*, 1990). Goodyear *et al.* (1990) reported a 4-fold increase in glucose transport by rat gastrocnemius muscle immediately following exercise (1.8-fold increase in transporter number, 1.9-fold increase in intrinsic activity) that declined to a 1.8-fold increase in transport after 30 min recovery (1.6-fold increase in transporter number, 1.1-fold increase in intrinsic activity; Goodyear *et al.*, 1990). Both transporter number and intrinsic activity had returned to basal levels by 2 h into the recovery period. Exercise and insulin stimulate muscle glucose uptake and glycogen synthesis in an independent and additive manner, supporting the existence of two separate pools of glucose transporters (Wallberg-Henriksson *et al.*, 1988; Sternlicht *et al.*, 1989; Goodyear *et al.*, 1990). Thus, insulin-independent and insulin-dependent recovery from glycogen-depleting exercise may be largely under the control of Glut4 transporters in these two pools. It is interesting that when glycogen is extensively depleted but the remaining glycogen concentration remains above 35 mmol/l, there is a rapid burst of increased glycogen synthesis that is quickly shut down (Price *et al.*, 2000). It is possible that this rapid burst of synthesis results from exercise-induced Glut4 stimulation that cannot be maintained because of feedback inhibition, again supporting local control by glycogen.

6.4.2. Insulin Dependence of Muscle Glycogen Recovery in Healthy Subjects

In our 1994 study of healthy fasting subjects insulin secretion was inhibited by an infusion of somatostatin, revealing that the initial rapid rate of recovery from glycogen depleting exercise is independent of insulin, while the slower subsequent rate is insulin dependent (Price *et al.*, 1994a). Glycogen synthesis was resumed 5 h into the recovery period, when the somatostatin infusion was discontinued and insulin levels returned to normal (Figure 6.2). The insulin independence of this rapid phase of glycogen synthesis, which had not been previously observed in non-diabetic human subjects, paralleled that of animal results (Richter *et al.*, 1984). Furthermore, the finding that in normal subjects insulin-dependent resynthesis also began at glycogen levels >30–40 mM indicated that the transition to insulin-dependent synthesis observed by Maehlum *et al.* (1977) was not related to the subjects being diabetic.

6.4.3. Muscle Glycogen Recovery in Insulin-resistant Subjects

Early diabetes studies in our laboratory demonstrated that glucose transport/phosphorylation is impaired under hyperglycemic–hyperinsulinemic conditions in both subjects with non-insulin-dependent diabetes mellitus (NIDDM) (Rothman *et al.*, 1992) and insulin-resistant (IR) offspring of subjects with NIDDM (Rothman *et al.*, 1995). The presence of this defect in IR offspring demonstrated that the reduction of muscle glucose transport/phosphorylation is fully expressed prior to the development of diabetes (Rothman

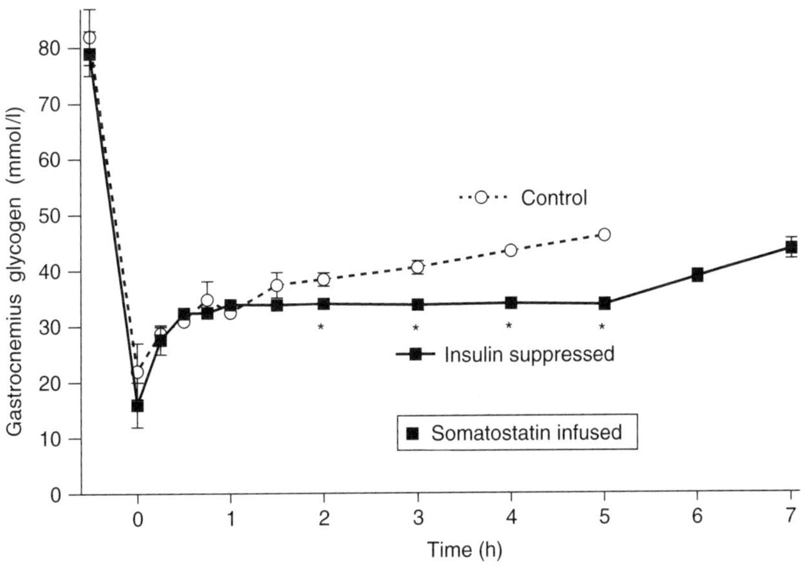

Figure 6.2. Mean glycogen (mmol/l) in gastrocnemius muscles of five subjects under normal insulin (○) and insulin-suppressed (■) conditions over 7 h fasting recovery. When somatostatin infusion was discontinued (5 h) glycogen synthesis resumed. $*p < 0.05$. Rates are means ± SE. (Reproduced from Price, T.B., Rothman, D.L., Taylor, R., Shulman, G.I., Avison, M.J. and Shulman, R.G., *J. Appl. Physiol.* **76**(1): 104–111, 1994 by permission of the American Physiological Society.)

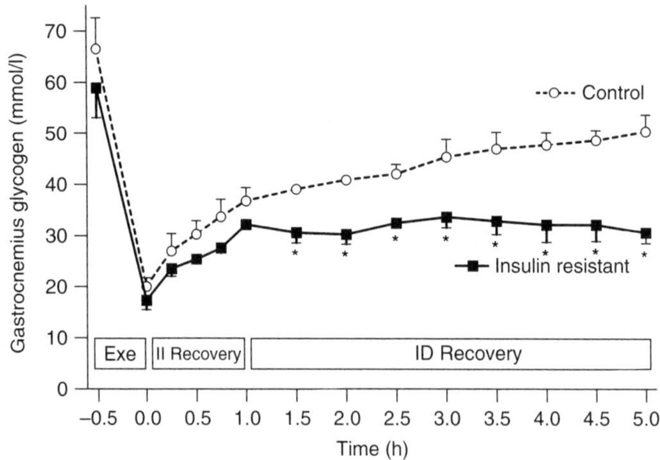

Figure 6.3. Mean glycogen (mmol/l) in gastrocnemius muscles of eight IR offspring (■) and eight healthy controls (○) over 5 h fasting recovery. $*p < 0.0016$ (control vs IR offspring). Rates are means ± SE. (Reproduced from Price, T.B., Perseghin, G., Duleba, A., Chen, W., Chase, J., Rothman, D.L., Shulman, R.G. and Shulman, G.I., *Proc. Natl Acad. Sci. USA* **93**: 5329–5334, 1996 by permission of the National Academy of Sciences.)

et al., 1995). In a 1996 exercise and recovery study, we reported that, in a group of IR offspring of parents with adult-onset diabetes, the insulin-independent phase of recovery from glycogen-depleting exercise was intact, while the insulin-dependent phase was impaired (Price *et al.*, 1996; Figure 6.3). The study compared age and weight matched healthy control subjects with otherwise healthy IR offspring. Plasma insulin levels in the IR group were two times greater ($p < 0.04$) than that of the controls, indicating that the impairment of insulin-dependent glycogen synthesis occurred in the presence of an abundance of insulin.

Recently we have reported that smoking also results in impairment of insulin-dependent recovery following glycogen-depleting exercise (Price *et al.*, 2003a). Habitual smoking is associated with insulin resistance (Oshida *et al.*, 1989) and impaired glucose tolerance in intravenous glucose tolerance tests (Frati *et al.*, 1996). Cigarette smoking alters metabolism at a number of different points that may impact insulin resistance. These include free fatty acid and glycerol mobilization (Colberg *et al.*, 1995), muscle glycogen utilization (Colberg *et al.*, 1994, 1995), circulating catecholamines (Colberg *et al.*, 1994), and blood lactate (Colberg *et al.*, 1995; Neese *et al.*, 1994) levels, as well as insulin-dependent glucose metabolism (Frati *et al.*, 1996; Jensen *et al.*, 1995). Insulin-resistant glucose metabolism is a well-established risk factor in the development of adult-onset diabetes (Frati *et al.*, 1996; Rothman *et al.*, 1992). People who smoke >15 cigarettes per day carry a two-fold greater risk of diabetes compared with those who have never smoked (Rimm *et al.*, 1995). Our 2003 study reported insulin-dependent glycogen synthesis rates that were different from controls and similar to IR offspring under the same post-exercise conditions (Price *et al.*, 1996, 2003a) (Figure 6.4). From these results it appears that moderate-to-heavy smokers suffer a similar reduction in insulin-dependent muscle glycogen synthesis. Thus in IR offspring, smoking may hasten the onset of NIDDM. Furthermore, smoking could increase the risk of developing NIDDM in those who are otherwise healthy with no family history of diabetes.

6.4.4. MRI to Locate Exercised Muscles

MRI can be used to measure muscle activity that appears, within groups of muscles, as signal increases that are workload dependent (Fisher *et al.*, 1991; Price *et al.*, 1995; Price and Gore, 1998; Sloniger *et al.*, 1998).

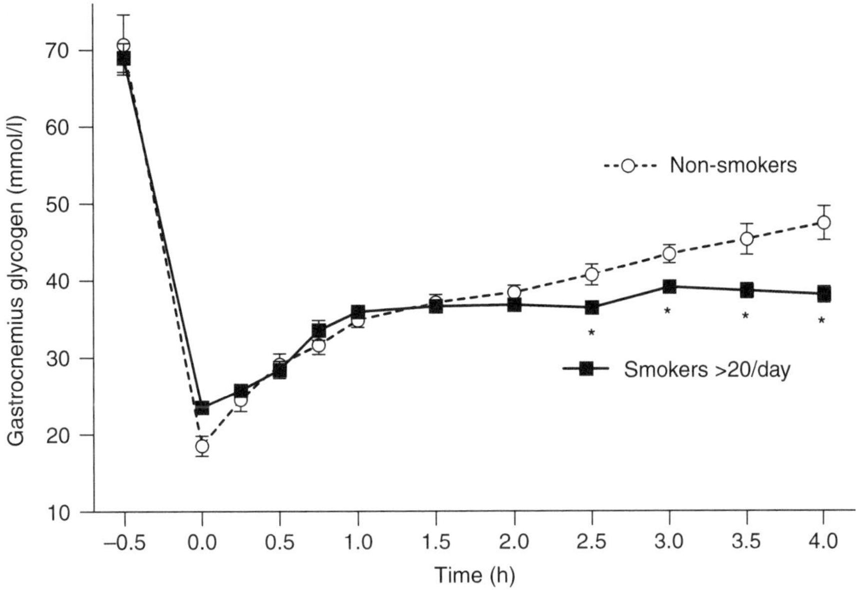

Figure 6.4. Mean glycogen (mmol/l) in gastrocnemius muscles of eight smokers (■) and 10 healthy controls (○) over 4 h fasting recovery. $*p < 0.05$ (control vs smokers). Rates are means ± SE. (Reproduced from Price, T.B., Krishnan-Sarin, S. and Rothman, D.L., *Am. J. Physiol. Endocrinol. Metab.* **285**(1): E116–122, 2003 by permission of the American Physiological Society.)

In contrast to non-exercising tissues, muscles recruited during exercise appear hyper-intense as MRI signal intensity increases with exercise. These exercise-induced signal changes result primarily from increases in the spin–spin (T_2) relaxation time of tissue water (Fleckenstein *et al.*, 1988; Fisher *et al.*, 1991; Price *et al.*, 1995) that are probably brought about by the metabolic changes associated with muscle contraction (Vandenborne *et al.*, 2000). MRI has been used to demonstrate activity in a number of muscles including: (1) the gastrocnemius, peroneus and soleus by plantar flexion (Fleckenstein *et al.*, 1988; Price *et al.*, 1995; Sloniger *et al.*, 1997; Price, 2003b); (2) dorsi-flexors (Fisher *et al.*, 1991; Price *et al.*, 1995); (3) gluteals, adductors, hamstrings and quadriceps (Sloniger, 1997); and (4) muscles of the forearm (Fleckenstein *et al.*, 1988). The magnitude of exercise-induced T_2 change is dependent upon workload and independent of individual muscle size (Fisher *et al.*, 1991; Price *et al.*, 1995). Because MRI detects similar behavior in all skeletal muscles with exercise and recovery, it is a valuable tool for identifying active muscles for further analysis with NMR spectroscopy.

Our glycogen depletion/repletion studies (Price *et al.*, 1991, 1994a, b, 1995, 1996, 2000, 2003a, b; Price and Gore, 1998) have relied on the assumption that, with the knee fully extended, the localized plantar flexion protocol activated only the gastrocnemius. With this assumption, we could accurately assess the workload, and we could think of the gastrocnemius as an isolated system. Numerous MRI studies have shown that plantar flexion with the knee fully extended activates the medial and lateral heads of the gastrocnemius with minimal contribution from the soleus (Fleckenstein *et al.*, 1988; Price *et al.*, 1995; Price and Gore, 1998; Vandenborne *et al.*, 2000). More recent studies have shown that muscle activation during plantar flexion is a shared phenomenon, with the degree of contribution of the plantar–flexor muscles being influenced by the degree of flexion of the knee joint (Sloniger *et al.*, 1997; Price *et al.*, 2003b), in agreement with EMG studies (Sale, 1982; Moritani and Muro, 1990; Bilodeau *et al.*, 1994; Tamaki *et al.*, 1997). Figure 6.5 shows MR images obtained before and after plantar flexion with the knee flexed [Figure 6.5(a, b)] and

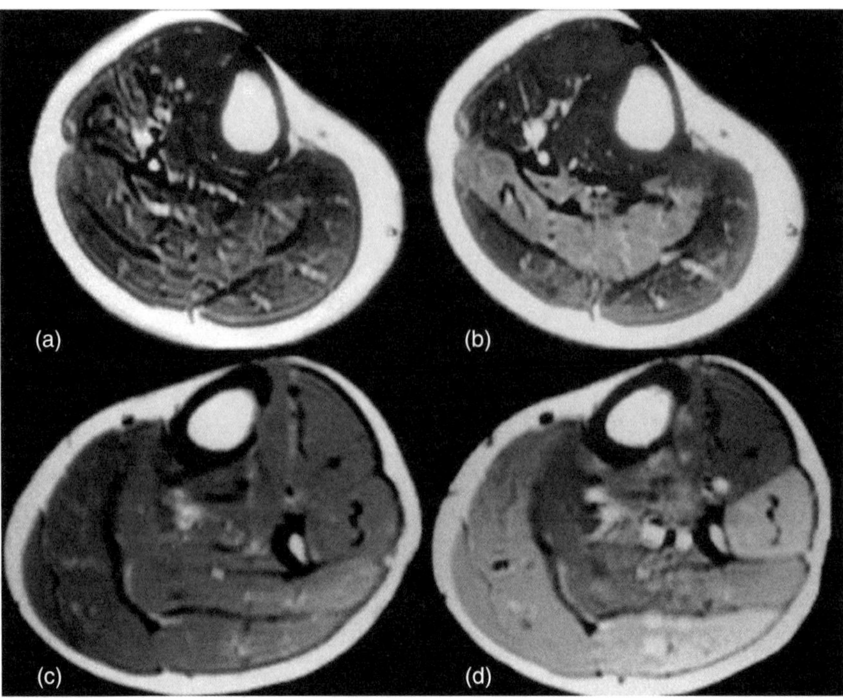

Figure 6.5. MRI of the lower leg (mid-calf) obtained before and immediately after exercise. (a) Before plantar flexion with the knee flexed at 90°; (b) after plantar flexion with the knee flexed at 90°; (c) before plantar flexion with the knee fully extended (0° flexion); (d) after plantar flexion with the knee fully extended (0° flexion). At full knee extension the preoneus was activated to stabilize the ankle. (Reproduced from Price, T.B., Kamen, G., Damen, B.M., Knight, C.A., Applegate, B., Gore, J.C., Eward, K. and Signorile, J.F., *Magn. Reson. Imag.* **21**: 853–861, 2003b by permission of Elsevier Science Ltd.)

with the knee extended [Figure 6.5(c, d)]. With the knee extended signal intensities are increased in the medial and lateral heads of the gastrocnemius, but not in the soleus. Signal intensity is also increased in the peroneus, indicating co-contraction of this muscle to stabilize the ankle.

A systemic study of the effects of fitness upon MRI, from a collaborative project between our laboratory and the *National Geographic*, is shown in Figure 6.6 (Gore, 2000). In this project a combination of non-invasive methods was used to compare performance in age-matched subjects at different levels of fitness. A fit subject who exercised regularly (Fit) was compared with an Olympic athlete (Ath) and a sedentary subject (Sed). These comparisons were made both at rest and following 1 h running on a treadmill at 75 % VO_{2MAX}. Left ventricular volumes (LVV) were measured from cardiac MRIs taken at rest. When compared with the fit subject, the LVV was increased by 12 % in the athlete while it was 22 % smaller in the sedentary subject. Arterial diameters (iliac and femoral), measured from MR angiograms taken at rest, revealed similar differences of 20–60 % increases in the athlete, and 7–40 % decreases in the sedentary subject.

The different colors in each participant's leg represent increases in transverse relaxation times (in ms) measured from MRIs obtained before and after running, with blue being less T_2 increase and yellow being greater T_2 increase. Although the athlete did 1.3 times the work of the fit subject and 2.3 times the work of the sedentary subject, his muscles' T_2 values did not change as much. In addition, of the total of 28 muscles studied in both the upper and lower body, the athlete significantly activated only 11 %, while the

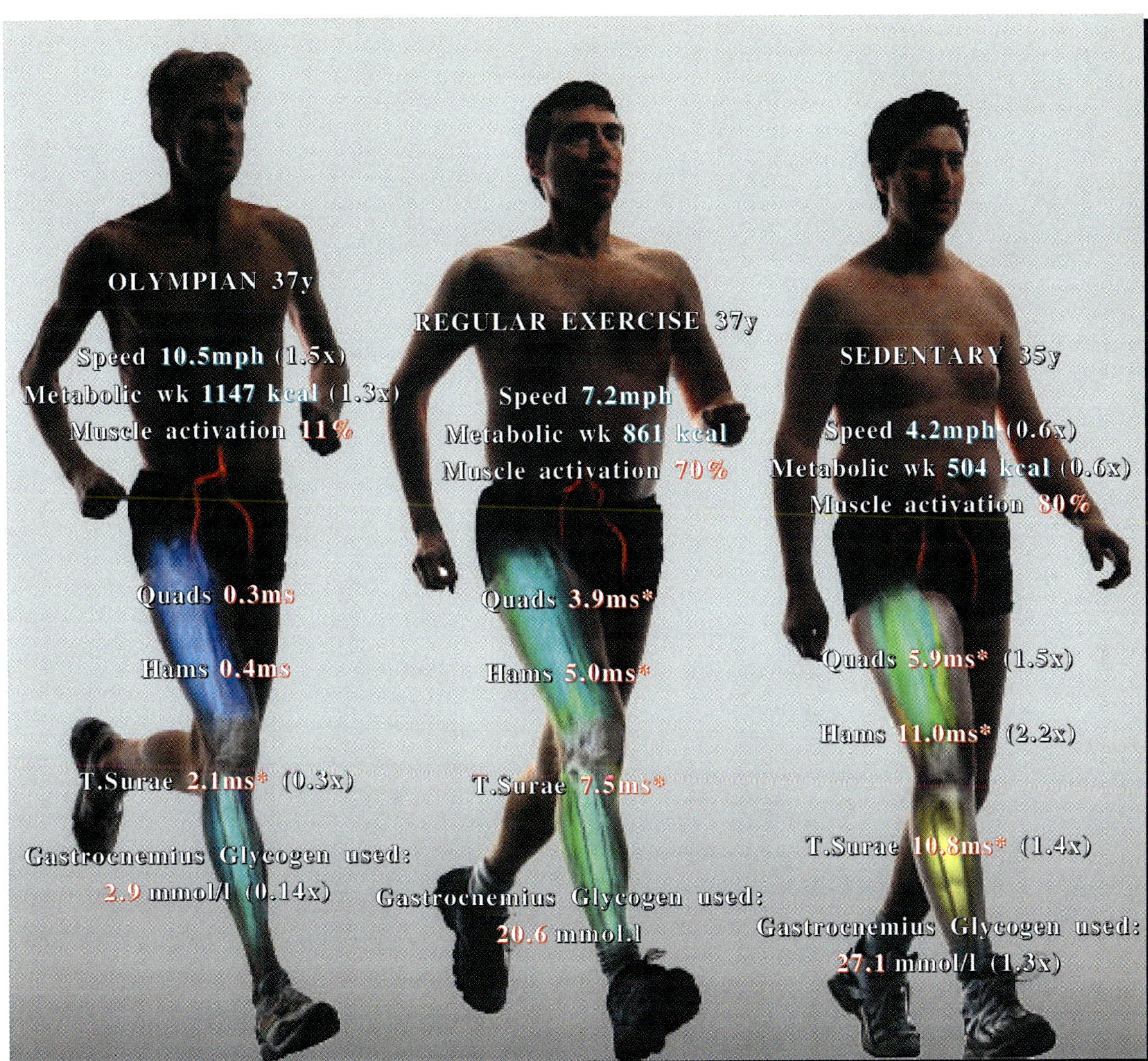

Plate 1 (Figure 6.6). Comparison of three subjects in their mid-thirties at different levels of fitness: an Olympic athlete (left), a fit individual who exercises regularly (three or four times per week) (middle), and a sedentary individual who does not exercise (right). Speed = running speed; numbers beside each muscle name = T_2 increase (ms) following exercise (* $p < 0.05$ vs at rest); gastrocnemius carbohydrate use numbers are calculated from NMR data. Muscle, bone, and vascular images are three-dimensional reconstructions of stacked axial MRIs.

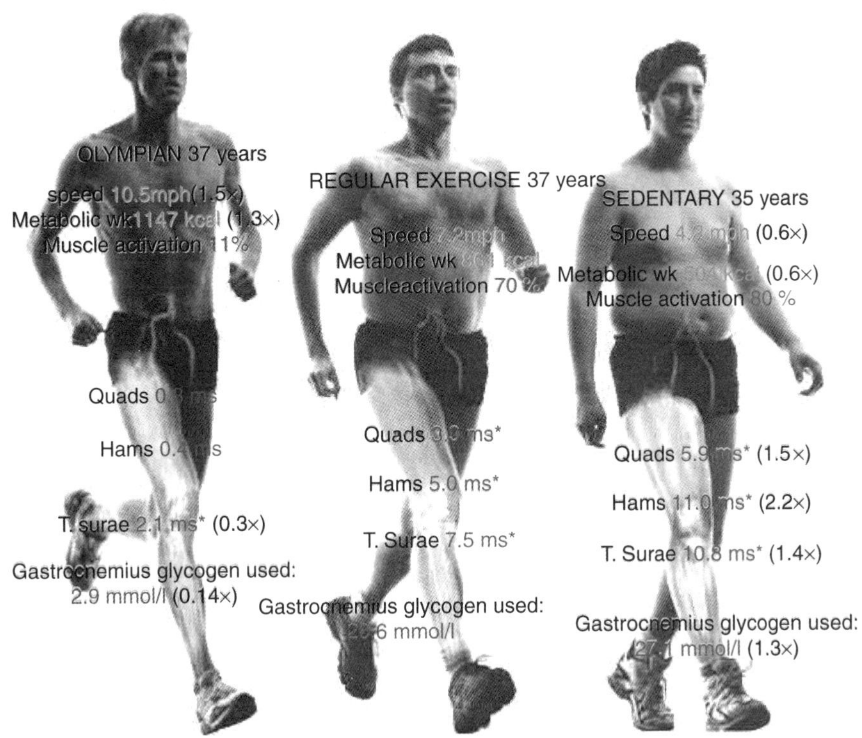

Figure 6.6 (Plate 1). Comparison of three subjects in their mid-thirties at different levels of fitness: an Olympic athlete (left), a fit individual who exercises regularly (three or four times per week) (middle), and a sedentary individual who does not exercise (right). Speed = running speed; numbers beside each muscle name = T_2 increase (ms) following exercise ($*p < 0.05$ vs at rest); gastrocnemius carbohydrate use numbers are calculated from NMR data. Muscle, bone and vascular images are three-dimensional reconstructions of stacked axial MRIs (see colour plate).

other subjects significantly activated 70–80 %. When MRI data were combined with ^{13}C NMR measurements of muscle glycogen use during the hour of running, the metabolic effect of fitness was revealed. The sedentary subject used 1.3 times more glycogen to do 0.6 times the amount of work as the fit subject, and the athlete used 0.14 times the amount of glycogen to do 1.3 times the work. While genetic variability is likely to contribute to these differences, the data clearly point to the benefits of regular physical activity as well as the risks of inactivity, and suggest that the benefit of heavy training may even be smaller than the detrimental effect of inactivity. In the future these combined technologies of MR imaging and spectroscopy will provide answers in the complex study of muscle metabolism that have heretofore been unattainable.

REFERENCES

Ahlborg, G., Wahren, J. and Felig, P. Splanchnic and peripheral glucose and lactate metabolism during and after prolonged arm exercise. *J. Clin. Invest.* **77**: 690–699, 1986.

Avison, M.J., Rothman, D.L., Nadel, E., Jue, T. and Shulman, R.G. Detection of human muscle glycogen by natural abundance ^{13}C NMR. *Proc. Natl Acad. Sci. USA* **85**: 1634–1636, 1988.

Bilodeau, M., Goulet, C., Nadeah, S., Arsenault, A.B. and Gravel, D. Comparison of the EMG power spectrum of the human soleus and gastrocnemius muscles. *Eur. J. Appl. Physiol.* **68**: 395–401, 1994.

Chance, B., Eleff, S., Leigh, J.S. Jr, Sokolow, D. and Sapega A. Mitochondrial regulation of phosphocreatine/inorganic phosphate ratios in exercising human muscle: a gated ^{31}P NMR study. *Proc. Natl Acad. Sci. USA* **78**(11): 6714–6718, 1981.

Cohen, S.M., Ogawa, S. and Shulman R.G. ^{13}C NMR studies of gluconeogenesis in rat liver cells: utilization of labeled glycerol by cells from euthyroid and hyperthyroid rats. *Proc. Natl Acad. Sci. USA.* **76**(4): 1603–1609, 1979.

Colberg, S.R., Casazza, G.A., Horning, M.A., and Brooks, G.A. Increased dependence on blood glucose in smokers during rest and sustained exercise. *J. Appl. Physiol.* **76**(1): 26–32, 1994.

Colberg, S.R., Casazza, G.A., Horning, M.A., and Brooks, G.A. Metabolite and hormonal response in smokers during rest and sustained exercise *Med. Sci. Sports Exercise* **27**(11): 1527–1534, 1995.

Dawson, M.J., Gadian, D.G. and Wilkie, D.R. Studies of the biochemistry of contracting and relaxing muscle by the use of ^{31}P n.m.r. in conjunction with other techniques. *Phil. Trans. R. Soc. Lond. Ser. B Biol. Sci.* **289**(1037): 445–455, 1980.

Dietrichson, P., Coakley, J., Smith, P.E.M., Griffiths, R.D., Helliwell, T.R. and Edwards, R.H.T. Conchotome and needle percutaneous biopsy of skeletal muscle. *J. Neurol. Neurosurg.* **50**: 1461, 1987.

Douen, A.G., Ramlal, T., Pastogi, S., Bilan, P.J., Cartee, G.D., Vranic, M., Holloszy, J.O. and Klip, A. Exercise induces recruitment of the 'insulin-responsive glucose transporter'. *J. Biol. Chem.* **265**(23): 13427–13430, 1990.

Fell, R.D., Terblanche, S.E., Ivy, J.L. and Holloszy, J.O. Effect of muscle glycogen content on glucose uptake following exercise. *J. Appl. Physiol.* **52**: 434–437, 1982.

Fisher, M.J., Meyer, R.A., Adams, G.R., Foley, J.M. and Potchen, E.J. Direct relationship between proton T_2 and exercise intensity in skeletal muscle MR images. *Invest. Radiol.* **25**(5): 480–485, 1991.

Fleckenstein, J.L., Canby, R.C., Parkey, R.W. and Peshock, R.M. Acute effects of exercise on MRI on skeletal muscle in normal volunteers. *Am. J. Roent.* **151**: 231–237, 1988.

Frati, A.C., Iniestra, F. and Ariza, C.E. Acute effect of cigarette smoking on glucose tolerance and other cardiovascular risk factors. *Diabetes Care* **19**(2): 112–117, 1996.

Gadian, D.G., Hoult, D.I., Radda, G.K., Seeley, P.J., Chance, B. and Barlow, C. Phosphorus nuclear magnetic resonance studies on normoxic and ischemic cardiac tissue. *Proc. Natl Acad. Sci. USA* **73**(12): 4446–4448, 1976.

Garetto, L.P., Richter, E.A., Goodman, M.N. and Ruderman, N.B. Enhanced muscle glucose metabolism after exercise in the rat: the two phases. *Am. J. Physiol.* **246**: E471–E475, 1984.

Gollnick, P.D., Piehl, K. and Saltin, B. Selective glycogen depletion pattern in human muscle fibres after exercise of varying intensity and at varying pedalling rates. *J. Physiol.* **241**: 45–57, 1974.

Goodyear, L.J., Hirshman, M.F., King, P.A., Horton, E.D., Thompson, C.M. and Horton, E.S. Skeletal muscle plasma membrane glucose transport and glucose transporters after exercise. *J. Appl. Physiol.* **68**(1): 193–198, 1990.

Gore, R. (Senior Editor), McNally, J. (photographer) and Price, T.B. (contributing author). What it takes to build the unbeatable body: pushing the limits, inner workings of fitness. *National Geographic* **198**(3): 14–15, 2000.

Griffin, J.H., Alazard, R., Dibello, C., Sala, E., Mermet-Bouvier, R. and Cohen, P. Carbon-13 nuclear magnetic resonance studies on (85 per cent ^{13}C-enriched Gly 9) oxytocin. *FEBS Lett.* **50**(2): 168–71, 1975.

Gruetter, R., Prolla, T.A. and Shulman, R.G. ^{13}C NMR visibility of rabbit muscle glycogen *in vivo*. *Magn. Reson. Med.* **20**: 327–332, 1991.

Harris, R.C., Hultman, E. and Nordesjo, L.-O. Glycogen, glycolytic intermediates and high-energy phosphates determined in biopsy samples of musculus quadriceps femoris of man at rest: methods and variance of values. *Scand. J. Clin. Lab. Invest.* **33**: 109–120, 1974.

Hermansen, L., Hultman, E. and Saltin, B. Muscle glycogen during prolonged severe exercise. *Acta Physiol. Scand.* **71**: 129–139, 1967.

Ivy, J.L. Muscle glycogen synthesis before and after exercise. *Sports Med.* **11**(1): 6–19, 1991.

Ivy, J.L., Chi, M.M.-Y., Hintz, C.S., Sherman, W.M., Hellendal, R.P. and Lowry, O.H. Progressive metabolic changes in individual human muscle fibers with increasing work rates. *Am. J. Physiol.* **252**: C630–C639, 1987.

Jensen, E.X., Fucsh, C., Jaeger, P., Peheim, E., and Horber, F.F. Impact of chronic cigarette smoking on body composition and fuel metabolism. *J. Clin. Endrocr. Metab.* **80**: 2161–2165, 1995.

Maehlum, S. and Hermansen, L. Muscle glycogen concentration during recovery after prolonged severe exercise in fasting subjects. *Scand. J. Clin. Lab. Invest.* **38**: 557–560, 1978.

Maehlum, S., Hostmark, A.T. and Hermansen, L. Synthesis of muscle glycogen during recovery after severe exercise in diabetic and non-diabetic subjects. *Scand. J. Clin. Lab. Invest.* **37**: 309–316, 1977.

McDermott, J.C., Elder, G.C.B. and Bonen, A. Non-exercising muscle metabolism during exercise. *Pflug. Arch.* **418**: 301–307, 1991.

Moritani, T. and Muro, M. Differences in modulation of the gastrocnemius and soleus H-reflexes during hopping in man. *Acta Physiol. Scand.* **138**: 575–576, 1990.

Neese, R.A., Benowitz, N.L., Hoh, R., Faix, D., LaBua, A., Pun, K. and Hellerstein, M.K. Metabolic interactions between surplus dietary energy intake and cigarette smoking or its cessation. *Am. J. Physiol. Endocrinol. Metab.* **267**(30): E1023–E1034, 1994.

Nielsen, H.V., Staberg, B., Nielsen, K. and Sejrsen, P. Effects of dynamic leg exercise on subcutaneous blood flow rate in the lower limb of man. *Acta Physiol. Scand.* **134**: 513–518, 1988.

Ogawa, S., Shulman, R.G., Glynn, P., Yamane, T. and Navon, G. On the measurement of pH in Escherichia coli by 31P nuclear magnetic resonance. *Biochim. Biophys. Acta* **502**(1): 45–50, 1978.

Oshida, Y., Yamanouchi, K., Hayamizu, S. and Sato, Y. Long-term mild jogging increases insulin action despite no influence on body mass index or $VO_{2\,max}$. *J. Appl. Physiol.* **66**: 2206–2210, 1989.

Price, T.B. and Gore, J.C. Effect of muscle glycogen content on exercise induced changes in muscle T_2 times. *J. Appl. Physiol.* **84**(4): 1178–1184, 1998.

Price, T.B., Rothman, D.L., Avison, M.J., Buonamico, P. and Shulman, R.G. ^{13}C NMR measurements of muscle glycogen during low-intensity exercise. *J. Appl. Physiol.* **70**(4): 1836–1844, 1991.

Price, T.B., Rothman, D.L., Taylor, R., Shulman, G.I., Avison, M.J. and Shulman, R.G. Human muscle glycogen resynthesis after exercise: insulin dependent and independent phases. *J. Appl. Physiol.* **76**(1): 104–111, 1994a.

Price, T.B., Taylor, R., Mason, G.M., Rothman, D.L., Shulman, G.I. and Shulman, R.G. Turnover of human muscle glycogen during low-intensity exercise. *Med. Sci. Sports Exercise* **26**(8): 983–991, 1994b.

Price, T.B., McCauley, T.R., Duleba, A.J., Wilkins, K.L. and Gore, J.C. Changes in magnetic resonance transverse relaxation times of two muscles following standardized exercise. *Med. Sci. Sports Exercise* **27**(10): 1421–1429, 1995.

Price, T.B., Perseghin, G., Duleba, A., Chen, W., Chase, J., Rothman, D.L., Shulman, R.G. and Shulman, G.I. NMR studies of muscle glycogen synthesis in insulin resistant offspring of NIDDM parents immediately following glycogen depleting exercise. *Proc. Natl Acad. Sci. USA* **93**: 5329–5334, 1996.

Price, T.B., Laurent, D., Petersen, K.F., Rothman, D.L. and Shulman, G.I. Glycogen loading alters muscle glycogen resynthesis after exercise. *J. Appl. Physiol.* **88**(2): 698–704, 2000.

Price, T.B., Krishnan-Sarin, S. and Rothman, D.L. Smoking impairs muscle recovery from exercise. *Am. J. Physiol. Endocrinol. Metab.* **285**(1): E116–122, 2003a.

Price, T.B., Kamen, G., Damen, B.M., Knight, C.A., Applegate, B., Gore, J.C., Eward, K. and Signorile, J.F. Comparison of MRI with EMG to study muscle activity associated with dynamic plantar flexion. *Magn. Reson. Imag.* **21**: 853–861, 2003b.

Richter, E.A., Garetto, L.P., Goodman, M.N. and Ruderman, N.B. Enhanced muscle glucose metabolism after exercise: modulation by local factors. *Am. J. Physiol.* **246**: E476–E482, 1984.

Rimm, E.B., Chan, J., Stampfer, M.J., Colditz, G.A. and Willett, W.C. Prospective study of cigarette smoking, alcohol use, and the risk of diabetes in men. *Br. Med. J.* **310**: 555–559, 1995.

Robergs, R.A., Pearson, D.R., Costill, D.L., Fink, W.J., Pascoe, D.D., Benedict, M.A., Lambert, C.P. and Zachweija, J.J. Muscle glycogenolysis during differing intensities of weight-resistance exercise. *J. Appl. Physiol.* **70**(4): 1700–1706, 1991.

Rothman, D.L., Shulman, R.G. and Shulman, G.I. ^{31}P NMR measurements of muscle glucose-6-phosphate: evidence for reduced insulin dependent muscle glucose transport or phosphorylation in non-insulin dependent diabetes. *J. Clin. Invest.* **89**: 1069–1075, 1992.

Rothman, D.L., Magnusson, I., Cline, G., Gerard, D., Kahn, C.R., Shulman, R.G. and Shulman, G.I. Decreased muscle glucose transport/phosphorylation is an early defect in the pathogenesis of non-insulin dependent diabetes mellitus. *Proc. Natl Acad. Sci. USA* **92**: 983–987, 1995.

Sale, D. Influence of joint position on ankle plantar flexion in humans. *J. Appl. Physiol. Respir. Environ. Exercise Physiol.* **52**: 1636–1642, 1982.

Saltin, B. and Karlsson, J. Muscle glycogen utilization during work of different intensities. *Adv. Exp. Med. Biol.* **11**: 289–300, 1971.

Shulman, G.I., Rothman, D.L., Chung, Y., Rossetti, L., Petit, W.A., Barrett, E.J. and Shulman, R.G. ^{13}C NMR studies of glycogen turnover in the perfused rat liver. *J. Biol. Chem.* **263**(11): 5027–5029, 1988.

Shulman, G.I., Rothman, D.L., Jue, T., Stein, P., DeFronzo, R.A. and Shulman, R.G. Quantitation of muscle glycogen synthesis in normal subjects and subjects with non-insulin dependent diabetes mellitus by ^{13}C nuclear magnetic resonance spectroscopy. *New Engl. J. Med.* **322**: 223–228, 1990.

Sillerud, G.I. and Shulman, R.G. Structure and metabolism of mammalian liver glycogen monitored by carbon-13 nuclear magnetic resonance. *Biochemistry* **22**: 1087–1094, 1983.

Sloniger, M.A., Crueton, K.J., Prior, B.M. and Evans, E.M. Lower-extremity muscle activation during horizontal and uphill running. *J. Appl. Physiol.* **83**(6): 2073–2079, 1997.

Sternlicht, E., Barnard, R.J. and Grimditch, G.K. Exercise and insulin stimulate skeletal muscle glucose transport through different mechanisms. *Am. J. Physiol.* **256**: E227–E230, 1989.

Tamaki, H., Kohji, K., Akamine, T., Sakou, T. and Kurata, H. Electromyogram patterns during plantar flexions at various angular velocities and knee angles in human triceps surae muscles. *Eur. J. Appl. Physiol.* **75**: 1–6, 1997.

Taylor, R., Price, T.B., Rothman, D.L., Shulman, R.G. and Shulman, G.I. Validation of ^{13}C NMR measurement of human skeletal muscle glycogen by direct biochemical assay of needle biopsy samples. *Magn. Reson. Med.* **27**: 13–20, 1992.

Taylor, R., Price, T.B., Katz, L.D., Shulman, R.G. and Shulman, G.I. Direct measurement of change in muscle glycogen concentration after a mixed meal in normal subjects. *Am. J. Physiol.* **265**: E224–E229, 1993.

Taylor, R., Magusson, I., Rothman, D.L, Cline, G.W, Caumo, A., Cobelli, C., and Shulman, G.I. Direct assessment of liver glycogen storage by ^{13}C nuclear magnetic resonance spectroscopy and regulation of glucose homeostasis after a mixed meal in normal subjects. *J. Clin. Invest.* **97**: 126–132, 1996.

Vandenborne, K., Walter, G., Ploutz-Snyder, L., Dudley, G., Elliott, M.A. and De Meirleir, K. Relationship between muscle T_2^* relaxation properties and metabolic state: a combined localized ^{31}P-spectroscopy and ^{1}H-imaging study. *Eur. J. Appl. Physiol.* **82**(1–2): 76–82, 2000.

Vollestad, N.K. and Blom, P.C.S. Effect of varying exercise intensity on glycogen depletion in human muscle fibres. *Acta Physiol. Scand.* **125**: 395–405, 1985.

Wallberg-Henriksson, H., Constable, S.H., Young, D.A. and Holloszy, J.O. Glucose transport into rat skeletal muscle: Interaction between exercise and insulin. *J. Appl. Physiol.* **65**(2): 909–913, 1988.

Zang, L.-H., Laughton, M.R., Rothman, D.L. and Shulman, R.G. ^{13}C NMR relaxation times of hepatic glycogen *in vitro* and *in vivo*. *Biochemistry* **29**: 6815–6820, 1990a.

Zang, L.-H., Rothman, D.L. and Shulman, R.G. ^{1}H NMR visibility of mammalian glycogen in solution. *Proc. Natl Acad. Sci.* **87**: 1678–1680, 1990b.

7

^{13}C NMR Studies of Heart Glycogen Metabolism

Maren R. Laughlin

National Institute of Diabetes and Digestive and Kidney Diseases, National Institutes of Health, Bethesda, MD 20892-5460, USA

Douglas L. Rothman and Robert G. Shulman

Yale University School of Medicine, New Haven, CT 06520-8043, USA

Metabolomics by In Vivo NMR. Edited by R. G. Shulman and D. L. Rothman
 ISBN: 0-470-84719-0

7.1. INTRODUCTION

The metabolism of the heart is dictated by its major function, which is to reliably and continuously pump adequate amounts of oxygenated blood to all tissues of the body where fuel, nutrients and waste products are exchanged. The required energy can be derived under most circumstances from oxidation of circulating fatty acids, lactate and glucose. Heart muscle will also oxidize available ketone bodies and acetate, and stored endogenous substrates in the form of triglyceride or glycogen can be mobilized during acute exercise or stress. Using ^{13}C NMR, it has been possible to assemble a picture of the dynamic behavior of myocardial glycogen stores in a variety of physiological states. This provides insight into the role and importance of glycogen for energy production and storage. These ^{13}C NMR studies also show that the regulation of myocardial glycogen metabolism differs from skeletal muscle and liver in ways which reflect the unique energetic requirements for sustaining heart function.

Cardiac glycogen metabolism shares many similarities with muscle glycogen metabolism. The glycogen polymer chain is elongated by addition of glucosyl units from uridine diphosphoglucose (UDPG) by glycogen synthase (GSase) and cleaved by glycogen phosphorylase (GP) in a process which adds inorganic phosphate to produce glucose-1-phosphate. The regulation of the activity of GSase and GP in heart is primarily through the action of hormones. Insulin deactivates GP and activates GSase, and epinephrine deactivates GSase and activates GP. These hormones act via signaling pathways which have recently been reviewed by Roach (1) and are described in more detail in Section 2.3 of this chapter. Under conditions of insulin infusion, the net rates of glycogen synthesis appear to depend both on the rate of glucose uptake and phosphorylation, and on the rate of entry of glycolytically produced pyruvate into the tricarboxylic acid–krebs–citric acid (TCA) cycle. This requires that the fluxes through GSase and GP respond to divert glucose carbons to storage when there is a mismatch between insulin-stimulated glucose uptake, energy demand and the availability of other substrates. This can be achieved in the absence of changes in GSase activity by small changes in the concentration of its activator G6P. Therefore, the results of studies using *in vivo* ^{13}C NMR to study insulin stimulated glycogen synthesis are consistent with the primary control of the glycogen synthesis flux being the relative fluxes of glucose transport/phosphorylation and glycolysis, as opposed to the activity of GSase.

The earliest NMR studies of heart employed ^{31}P NMR to monitor high energy phosphate metabolism in perfused heart and *in vivo* (2, 3). The first ^{13}C NMR studies of heart glycogen were done in anesthetized, open-chested guinea pigs by Neurohr *et al.* (4, 5) in a horizontal-bore TMR-32/200 Oxford Research Systems magnet at Yale University. A solenoid coil tuned to the carbon frequency (20.19 MHz), surrounded by a saddle-shaped proton decoupling coil, was positioned within the chest cavity around the entire heart. Because of the low natural abundance of ^{13}C, glycogen was not visible in the unlabeled ^{1}H-decoupled ^{13}C NMR spectra of heart. Plasma glucose is the major substrate for new glycogen synthesis because the heart, like muscle, lacks the enzymes of gluconeogenesis. Therefore, when [1-^{13}C]glucose and insulin were infused into the jugular vein, the resultant peak at 100.6 ppm in the ^{13}C NMR spectrum was identified as [1-^{13}C]glycogen by its chemical shift, and by its expected increase with insulin and decrease in anoxia. These spectra were similar to those of *in vivo* rat heart shown in Figure 7.1. These early experiments

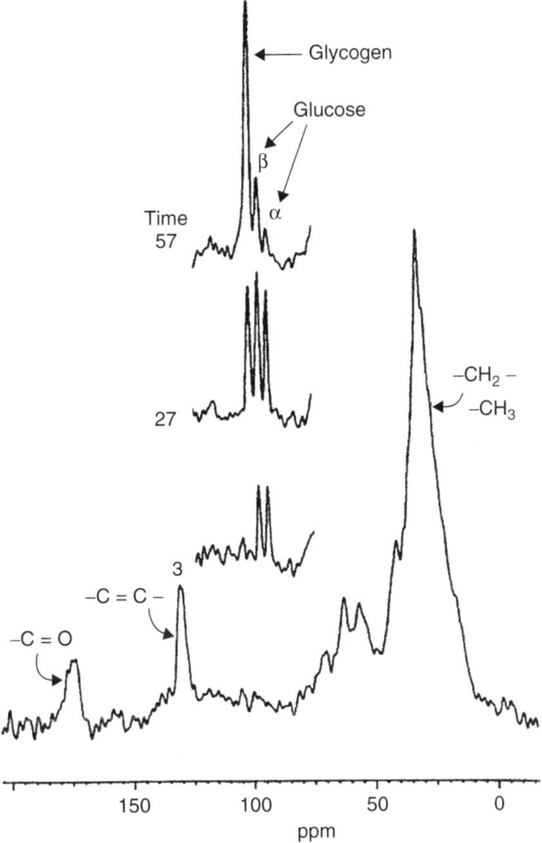

Figure 7.1. A natural abundance, proton-decoupled ^{13}C NMR spectrum of a rat heart obtained *in vivo* at 50.4 MHz. Peaks at 175, 130 and 30 ppm correspond to the carbonyl, double-bonded and single-bonded carbons of tissue fats. Inset: spectra taken at 3, 27 and 57 min after beginning a 50 min, 10 mg/min [1-^{13}C]glucose and 1 U/min insulin infusion. Peaks at 96.6 and 92.7 ppm are β- and α-[1-^{13}C]glucose, respectively, whereas the peak at 100.6 ppm arises from newly synthesized [1-^{13}C]glycogen. (Reproduced from Laughlin MR, Petit WA Jr, Shulman RG, Barrett EJ, *Am J Physiol* 1990; **258**(1 Pt 1): E184–E190 by permission of the American Physiological Society.)

demonstrated the value of using ^{13}C NMR to study cardiac carbohydrate metabolism in living animals, where the heart has access to its normal substrates, and can interact with the other systems in the body, so that the metabolic consequences of the trauma of heart isolation and perfusion are avoided.

In this chapter we will focus on ^{13}C NMR studies of cardiac glycogen synthesis. In keeping with the theme of this book we will examine whether an metabolic control analysis (MCA)-based model of heart glycogen metabolism, similar to that developed for skeletal muscle (Chapters 4 and 5) but differing based on the unique metabolic needs of the heart, can explain the extensive ^{13}C NMR data which exist on the response of heart glycogen metabolism to hormones and substrates. The chapter is organized in four sections. Section 7.2 will review the studies which established the NMR methodology to study cardiac glycogen metabolism in perfused heart and in the intact animal. The goal of this section is to provide the reader with the background to assess the strengths and limitations of the NMR measurement. Section 7.3 reviews ^{13}C NMR studies of the effects of hormones and substrates on glycogen synthesis, and section 7.4

discusses studies in diabetes and fasting, and states where myocardial glycogen metabolism is altered. In Section 7.5, a MCA model of heart glycogen synthesis is proposed and shown to be consistent with both the results reviewed in Sections 7.3 and 7.4, and results from recent studies in transgenic mice. The control is shown to be similar to muscle in that GSase has a low control coefficient (see Chapter 3) and therefore the rate of glycogen synthesis is largely determined by the rates of glucose transport/phosphorylation and glycolysis. However the control differs due to the much greater importance of glycolysis and the ability of the heart to oxidize other substrates such as lactate and ketones. It is our hope that this model will provide the impetus for further experimentation in the field in order to develop a quantitative understanding of the control of heart glycogen synthesis.

7.2. ^{13}C NMR METHODOLOGY FOR STUDIES OF HEART GLYCOGEN

7.2.1. Relaxation Properties and NMR Visibility of the Heart Glycogen Molecule

NMR is a quantitative technology, and can provide accurate measures of metabolic flux if the relationship between the observed signal and the metabolite concentration is known. It must therefore be asked whether the signal in the NMR spectrum reports on the total ^{13}C carbon present in glycogen. The answer to this question depends on the physical nature and environment of the molecules. Is the relaxation time (T_2*) long enough to yield sharp peaks? Are the physical structure and location of the glycogen pool homogeneous? Tissue glycogen acts chemically and physically as though it were in solution because of the large number of associated water molecules (6). Heart glycogen particles have a molecular weight of 10^7-10^9. They are associated with a variety of proteins, including a core protein glycogenin, enzymes GSase, GP, branching and debranching enzymes, protein kinases and phosphatases that alter the activity of GSase and GP, and the regulating proteins that modulate the activity of the kinases and phosphatases (1). Total glycogen can be extracted from tissue with hot alkali, but only partially extracted with acid. The protein-rich acid-precipitated glycogen fraction tends to be lower in molecular weight than that found in the supernatant, indicating that it may be less mature glycogen particles (7). Given these different fractions and the large, highly structured nature of the glycogen particle, it was necessary to determine whether heart glycogen could be accurately measured in tissues with NMR. Liver glycogen, which exists in much larger particles than found in heart, had been shown to be completely visible in the ^{13}C NMR spectrum, despite its very large molecular weight and size (Chapter 2). This was done both by calculating the expected peak shapes from the relaxation times measured *in vivo*, and by comparing the fully relaxed signal of extracted glycogen with the glucose resulting from complete hydrolysis (6, 8). In intact heart at 20.19 MHz, the T_1 for carbon 1 of glycogen was 48 ± 5 ms, and the nuclear Overhauser effect (NOE) was 1.3 ± 0.05. Similar to liver, fully relaxed signal intensities were the same in extracted heart glycogen and the glucose that resulted from complete hydrolysis (5). This indicated that purified heart glycogen was likely to be 100 % visible in the spectrum, similar to liver glycogen.

Garlick and Pritchard evaluated the visibility of glycogen in perfused rat heart at 100.6 MHz. Basal, unlabeled glycogen was depleted by injecting the rats prior to the experiment with isoproterenol to activate GP, and [1-^{13}C]glucose was added to the perfusate to completely label the glycogen pool as it was resynthesized. Fully relaxed proton-coupled and partially saturated decoupled spectra were acquired. [1-^{13}C]glycogen was quantified by correcting the signal area for saturation, filter effects, filling factor and NOE, and by calibration with a standard solution of [6-^{13}C]glucose. The ratio of calculated [1-^{13}C]glycogen from the NMR spectra of perfused heart to measured extracted [1-^{13}C]glycogen was between 1.01 and 1.09; glycogen in the intact heart appears to be completely visible in the NMR spectrum (9). These simple observations made it possible to calculate the net fluxes in or out of the cardiac glycogen pool using time dependent NMR experiments.

7.2.2. Quantitation of Myocardial Glycogen Synthesis

7.2.2.1. Perfused heart

^{13}C NMR provides a means to monitor the time dependence of changes in ^{13}C-labeled glycogen signal. The glycogen signal intensity in perfused hearts can be converted to a quantitative measure of glycogen concentration within limits as noted above (9), and the rate of glycogen synthesis can be calculated from the rate of change of signal in time. This approach was taken by Hoekenga *et al.* (10) to measure the effects of elevated glucose and potassium on the rates of glycogen synthesis and ischemic degradation in perfused guinea pig heart. Isolated guinea pig hearts were perfused with [1-^{13}C]glucose and insulin in a 20 mm NMR tube in the magnet. The rate of [1-^{13}C]glycogen synthesis during 40 min of normal flow was about 3 μmol/min g dry weight with low (3.0 mM) glucose and insulin, and was not affected by elevated glucose (11.7 mM), or the following 15 min of ischemia.

7.2.2.2. In vivo studies

Several of the assumptions and measurements easily made in perfused heart studies are not as easy to implement for experiments performed in intact animals. The filling factor coil loading and shimming can be variable due to differences in the size of the heart and chest cavity, and to motion of the lungs and heart. The ^{13}C-labeled glucose substrate for glycogen synthesis must be infused into the bloodstream, where it mixes with endogenous plasma glucose. Therefore, the enrichment of the substrate for glycogen is less than 100 %, and must be measured. These problems were addressed in rat (11) and dog (12, 13). Rats were intubated and ventilated, and catheters were placed in the jugular vein for substrate infusion and the carotid artery for blood sampling. The chest was opened and a double-tuned ^{13}C–^{1}H solenoid receiver coil was placed around the heart. [1-^{13}C]glycogen accumulation was monitored in NMR spectra during infusion of insulin and [1-^{13}C]glucose. Arterial blood samples were drawn every 15 min for measurement of [1-^{13}C]glucose enrichment. At the end of the experiment, the hearts were immediately removed and freeze-clamped for measurement of total and [1-^{13}C]glycogen content. Plasma [1-^{13}C]glucose enrichment was quite high, 30–50 %, and was measured using ^{1}H NMR. Figure 7.2 shows the singlet of the proton on C1 of unlabeled glucose at 5.27 ppm, and the doublet (5.50 and 5.05 ppm) from the proton attached to the

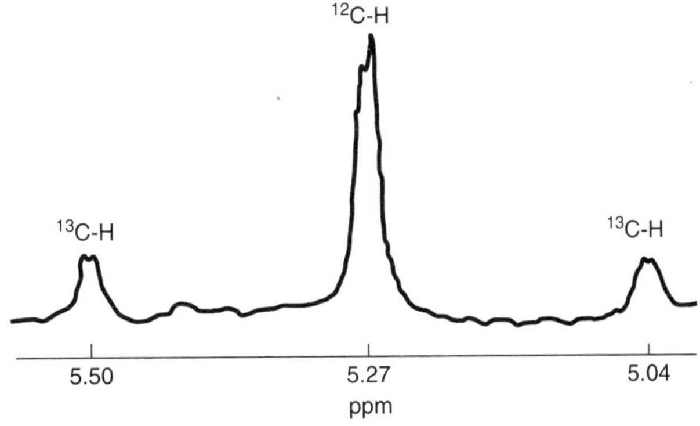

Figure 7.2. ^{1}H NMR spectrum taken at 360.1 MHz of the α-C1 proton of rat plasma glucose after 30 min of [1-^{13}C]glucose and insulin infusion. The area of the central peak is proportional to unlabeled glucose and the sum of the areas of the two satellites is proportional to [1-^{13}C]glucose. (Reproduced from Laughlin MR, Petit WA Jr, Dizon JM, Shulman RG, Barrett EJ, *J Biol Chem* 1988; **263**(5): 2285–2291 by permission of the American Society for Biochemistry and Molecular Biology.)

same carbon of [1-^{13}C]glucose. The ratio of peak areas (A) yields [1-^{13}C]glucose enrichment ($E_{glucose}$):

$$E_{glucose} = \frac{[1\text{-}^{13}\text{C}]\text{glucose}}{\text{Total glucose}} = \frac{A_{(5.5\,\text{ppm})} + A_{(5.05\,\text{ppm})}}{A_{(5.5\,\text{ppm})} + A_{(5.27\,\text{ppm})} + A_{(5.05\,\text{ppm})}}$$

Glycogen was extracted from the hearts, hydrolyzed, and total glucose was measured enzymatically. The ^{13}C enrichment at C1 of the resultant glucose could be measured by ^{1}H NMR as in Figure 7.2, but this proved difficult for the low enrichments seen (6–21 %). However, the C1 resonance could be compared to C2–$C5$ in high resolution, fully relaxed ^{13}C spectra of extracted hydrolyzed glycogen, where C2–C5 signals were assumed to represent the natural abundance ^{13}C enrichment of 1.1 %. The ^{13}C glycogen in hearts was therefore calculated from the following equation and represents the glycogen signal in the final ^{13}C NMR spectrum.

$$[1\text{-}^{13}\text{C}]\text{glycogen}_{final} = \text{total glycogen} \times E_{glycogen}$$

Finally, glycogen synthesis rate (Flux$_{glycogen}$) was calculated from the change in glycogen signal area over time ($A_{glycogen}$):

$$\text{Flux}_{glycogen} = \frac{\Delta A_{glycogen}}{\Delta \text{time}} \times \frac{[1\text{-}^{13}\text{C}]\text{glycogen}_{final}}{A_{glycogen,final}} \times \frac{1}{E_{glucose}}$$

The net rate of glycogen synthesis was constant throughout the experiment at $0.23 \pm 0.10\,\mu$mol/min g wet wt, very similar to the GSase enzyme activity measured at physiological substrate and activator concentrations, 0.21 μmol/min g wet weight (11). Net synthetic flux is the difference between the absolute rate of synthesis and the rate of breakdown in the observed, labeled layer of glycogen. This 'turnover' rate had to be measured before it was possible to understand the concordance between the measured flux through a pathway *in vivo* and the activity of the putative rate-limiting enzyme measured in tissue extracts, and the consequences for regulation of the glycogen pathway.

7.2.3. Ordered Synthesis and Breakdown – Measuring Glycogen Turnover *In Vivo*

The glycogen molecule is unique in that it has an ordered three-dimensional structure and interpretation of ^{13}C NMR experiments is dependent on the nature of the interplay between structure and metabolism. At the center of the 15–30 nm spherical molecule is a self-glycosylating protein glycogenin (1). Glucosyl units are added via α[1–4] linkages in a polymer chain. An α[1–6] branch point is found every eight to 12 residues, giving rise to new chains. Each chain gives rise on average to two new chains, and so each concentric tier of the glycogen particle has on average twice as much glycogen as the one below it, twice as many glucose chains, and twice as many non-reducing ends that can serve as substrates for GSase or GP. As glycogen is synthesized, glucose is added to the outside of the particle by GSase, which is associated with the surface of the particle. When it is hydrolyzed, glucosyl units are removed from the outside the particle by the associated GP protein. When [1-^{13}C]glucose is added, all newly synthesized ^{13}C-glycogen will have the same enrichment as its glucose precursor. If glycogen is breaking down at the same time as it is synthesized (turnover), a question arises – will only labeled, observable glycogen be lost, or will hydrolysis also occur in older glycogen from pre-existing, completed particles that were never labeled?

Studies were conducted in perfused heart (14) and in intact rats (5, 11, 15) to investigate glycogen turnover. A 'pulse-chase' type of experiment was used where ^{13}C-glucose labeled in one carbon was used as a substrate under conditions (high glucose and insulin) that support glycogen synthesis. After a period of time, the substrate was switched either to unlabeled glucose (5, 15) or to glucose labeled at a different carbon (14). In Figure 7.3, it can be seen that synthesis of labeled glycogen is linear with a constant slope during the first and last hours of a 3 h experiment, indicating that the net rate of glycogen synthesis is constant throughout. The labeled glycogen falls about 20 % over the second hour of the experiment while

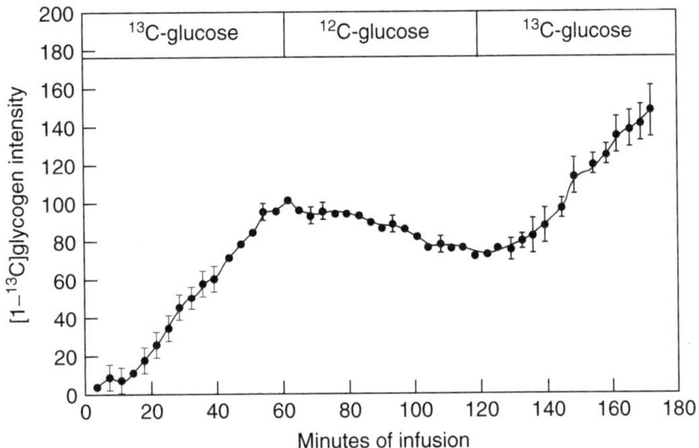

Figure 7.3. The intensity of the [1-^{13}C]glycogen signal in a 'pulse-chase' experiment in rat heart *in vivo*. The infusion consisted of 1 U/min insulin and 10 mg/min of [1-^{13}C]glucose (0–60 and 120–180 min) or unlabelled glucose (60–120 min). The intensities of the [1-^{13}C]glycogen signals were normalized to 100 % at 60 min. (Reproduced from Laughlin MR, Petit WA Jr, Barrett EJ, *J Mol Cell Cardiol* 1993; **25**(2): 175–183 by permission of Elsevier Science Ltd.)

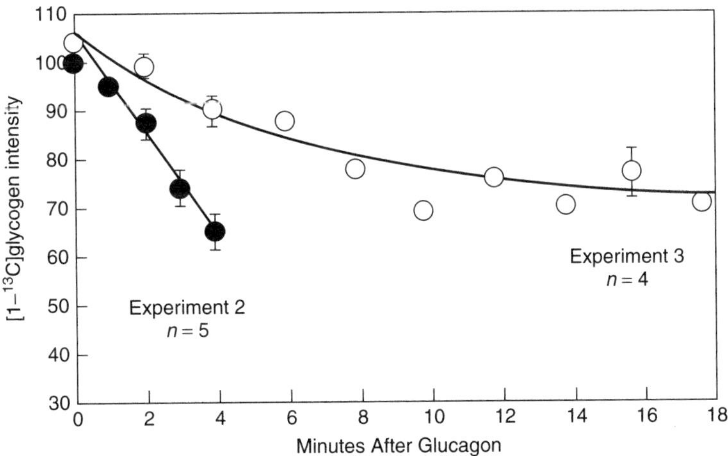

Figure 7.4. The intensity of the [1-^{13}C]glycogen signal in rat heart taken after a glucagon bolus. In experiment 2 (solid circles), glucagon was given after a 50 min infusion of [1-^{13}C]glucose (10 mg/min) and 1 U/min insulin, followed by 10 min saline. In experiment 3 (open circles), rats received the same infusions of labeled glucose and saline, followed by 50 min of unlabeled glucose and insulin, and 10 min of saline before the glucagon bolus. Intensities were normalized to 100 % for the largest glycogen resonance just before the glucagon bolus in each experiment. (Reproduced from Laughlin MR, Petit WA Jr, Barrett EJ, *J Mol Cell Cardiol* 1993; **25**(2): 175–183 by permission of Elsevier Science Ltd.)

unlabeled glycogen is synthesized (15). Therefore, some liberation of glucose from glycogen occurs even during this period of net synthesis, and only a lower limit estimation of the rates of hydrolysis and synthesis can be made. The action of GP in separate layers can be seen clearly during a period of net glycogenolysis initiated by glucagon in an intact rat after a similar 'pulse-chase' type of experiment (Figure 7.4). The most recently synthesized layer (solid circles) was liberated five times faster than the glycogen synthesized earlier

[2.5 ± 0.7 vs 0.5 ± 0.1 μmol/min g wet weight (15)]. Therefore, glycogen in the heart behaves as though it is built up and liberated mostly in layers, but not rigidly so. These experiments imply that metabolically active heart glycogen exists mostly in particles that are made in consecutive tiers and are broken down mostly in a 'last in, first out' manner.

The measured net synthesis rate therefore underestimates the actual flux through GSase by as much as 20 % due to simultaneous glycogenolysis. The measured GSase flux and activity, shown above, are similar within the accuracy of the NMR measurement, but the *in vivo* net synthesis rate must actually be slightly lower than the activity of the enzyme measured at appropriate concentrations of the substrates and activators.

7.3. REGULATION OF GLYCOGEN SYNTHESIS BY HORMONES AND METABOLITES

Insulin is the primary hormone which activates glycogen synthesis in the heart, whereas epinephrine is thought to be a major signal for degradation. The complex signaling pathways for insulin and epinephrine allow heart glycogen metabolism to be highly responsive to physiological inputs to the heart. In addition to hormonal control both GP and GSase have their own primary metabolite modulators. Glucose-6-phosphate (G6P) is a signal that glucose is abundant, and also allosterically activates GSase and glycogen storage. Adenosine 5′-monophosphate (AMP) is a signal that cellular energy stores are low, and also binds to and activates GP. GP is inhibited by adenosine triphosphate (ATP) and G6P, which are abundant when energy stores are high. Therefore, GP and GSase activity can both be very sensitive to the energy state and needs of the cell. As shown below, the effect on heart glycogen metabolism of nutrients such as glucose, lactate, ketones, and fatty acids is likely to be primarily via these allosteric effectors.

The complex interplay between hormones and metabolites in regulating cardiac glycogen synthesis has led to some uncertainty as to the relative physiological importance of each factor, as well as seemingly paradoxical results in the literature. The ability of ^{13}C NMR to study glycogen synthesis in the intact animal has made important contributions to understanding the interplay between these factors in the physiological state. In this section we review these studies, and in Section 7.5 further interpret them using a MCA based model to form the framework for quantitatively understanding substrate and hormonal regulation of cardiac glycogen synthesis.

7.3.1. *In vivo* ^{13}C NMR Studies of the Effects of Insulin and Epinephrine on Heart Glycogen Metabolism

7.3.1.1. Insulin

It has long been understood that insulin and epinephrine are major hormonal effectors of the enzymes involved in heart glycogen metabolism. Insulin stimulates glycogen synthesis in the heart by recruiting glucose transporter-4 (GLUT-4) to the plasma membrane and increasing glucose uptake and phosphorylation, and by the dephosphorylation (via a glycogen synthase phosphatase) and activation of GSase (1). The resultant accelerated glycogen synthesis can be readily observed in ^{13}C NMR experiments (Figures 7.1 and 7.3). Studies looking at how epinephrine and nutrients modulate glycogen synthesis under high insulin conditions are described below.

7.3.1.2. Epinephrine

Epinephrine (released during exercise, hypoxia and ischemia) and the counter-regulatory hormone glucagon (Figure 7.4) cause activation of cAMP-dependent protein kinases, which activate glycogen synthase kinases and deactivate GSase. Many of the same regulatory protein kinases and phosphatases use GP as a substrate.

7.3.1.3. Insulin and epinephrine

In working rat hearts perfused with glucose, preformed [1-^{13}C] glycogen is most rapidly released for the first 5–10 min or so of epinephrine stimulation, and this is followed by an increased uptake of glucose from the perfusate (16). The excess energy production needed for increased work under these conditions is derived solely from the oxidation of endogenous and exogenous carbohydrate (17). In anesthetized, open-chest dog, where glucose uptake was already elevated with exogenous insulin, very little breakdown of [1-^{13}C]glycogen was observed during a 10 min epinephrine infusion despite the elevation of mean arterial pressure from 110 ± 14 to 186 ± 21 mmHg and heart rate from 117 ± 3 to $175 \pm 17\ \text{min}^{-1}$ (Figure 7.5). However, following the epinephrine challenge, glycogen synthesis was stimulated 2-fold over the basal rate, despite the fact that glucose and oxygen uptake were the same before and after epinephrine (12). It would appear that the hearts in the intact anesthetized dog were able to increase their heart work without resort to mobilization of endogenously stored carbohydrate, implying that sufficient other nutrients were available for energy production. Moreover, despite no change in glucose uptake these hearts increased storage and decreased glucose oxidation as soon as the epinephrine challenge was over. What could allow for this phenomenon?

The lactate uptake by the heart, which is a function of arterial lactate concentration, was increased from 0.33 ± 0.08 to 0.72 ± 0.2 μmol/min g wet weight after exposure to epinephrine, when glucose uptake was unchanged (12). A similar phenomenon occurred following repeated short (3 × 90 s) periods of global hypoxia in the intact rat. Myocardial GSase was activated following hypoxia in healthy rats infused with insulin, and this was accompanied by an increased rate of glycogen synthesis as measured with ^{13}C NMR (0.45 ± 0.05 μmol/min g after hypoxia vs 0.32 ± 0.04 before hypoxia) (19). Hypoxia, like epinephrine infusion, is accompanied by large increases in plasma lactate, a nutrient which is readily taken up and oxidized by the heart. Therefore, lactate utilization does not appear to interfere with glucose uptake, but it does appear to route it from oxidation toward excess glycogen synthesis. Additional evidence supporting this idea was obtained in the isolated heart, where the perfusate milieu was altered to more resemble that found *in vivo* during exercise by including glucose, oleate, lactate and insulin. Under these conditions, epinephrine

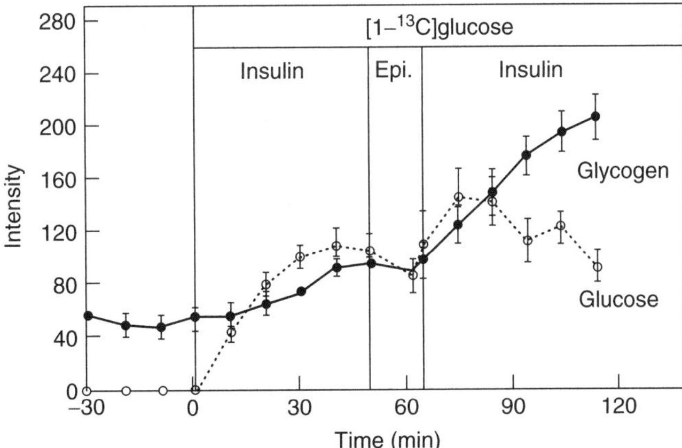

Figure 7.5. Intensities of beta[1-^{13}C]glucose (open circles) and [1-^{13}C]glycogen (solid circles) in 10 min ^{13}C NMR spectra of dog heart *in vivo*. Epinephrine was given at 50 min, and insulin was resumed 15 min later. Intensities were normalized so that glycogen was set to 100 % in the last spectrum before epinephrine was given. Most dogs possessed visible cardiac glycogen in the basal state, as reflected in the figure. (Reproduced from Laughlin MR, Taylor JF, Chesnick AS, Balaban RS, *Am J Physiol* 1992; 262(6 Pt 1): E875–883 by permission of the American Physiological Society.)

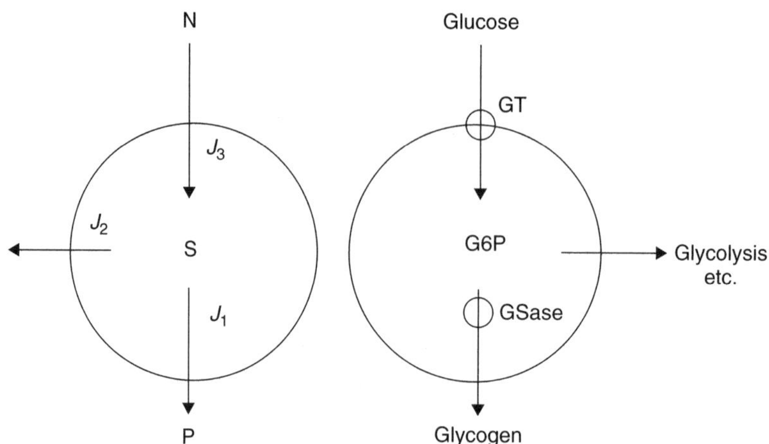

Figure 7.6. Model of coordinated control proposed by Kascer and Acarenza to accommodate changes in flux from N to the product (P) while maintaining the concentration of the shared intermediate (S) constant. In this way the branching flux J_2 will be independent of J_3. The analogy with the glycogenic pathway is shown where the shared intermediate is G6P.

caused increased heart activity and lactate and oleate oxidation, but did not stimulate glycogenolysis (18). Together these results imply a protective effect of exercise- or hypoxia-generated circulating lactate (from muscle) on heart glycogen stores, which may have the effect of increasing endurance. We describe below the results of studies looking directly at the effect of lactate on heart glycogen synthesis.

7.3.1.4. Effects of nutrients on glycogen synthesis

Plasma glucose is the direct precursor for heart glycogen synthesis. Unlike muscle, wherein the presence of insulin glycogen synthesis is linearly related to plasma glucose levels (to over 20 mM) (20), there is little increase in heart glucose phosphorylation and glycogen synthesis with hyperglycemia. This is consistent with the idea that the insulinized heart has a low K_m for glucose transport. Despite this, glucose is not a major substrate for heart energy production under most conditions; most of its energy needs are met with fatty acid oxidation, and it can also use pyruvate, lactate and ketone bodies. This was explored further using ^{13}C NMR in open-chested dogs. [1-^{13}C] glucose was infused, and the synthesis of [1-^{13}C]glycogen was monitored as described above. When unlabeled lactate, pyruvate or beta-hydroxybutyrate were infused peripherally to raise circulating concentrations of these alternative substrates, plasma glucose was unaltered. The rate of heart [1-^{13}C]glycogen arising from plasma [1-^{13}C]glucose was dramatically increased 5-fold in the presence of all exogenous nutrients (13). These nutrients were all oxidized by the heart in preference to plasma glucose or glycogen. [3-^{13}C]lactate is not observed in spectra of the heart, indicating that little myocardial glucose is disposed of as lactate. As discussed in Section 7.4, the observed increase in glycogen synthesis is therefore most likely due to the inhibition of glucose oxidation and glycolysis by competition from other nutrients, without coordinated inhibition of glucose transport and phosphorylation. Excess glucose is diverted through GSase, probably stimulated by increased substrate (UDPG) and activator (G6P) concentrations. Therefore, glycogen synthesis flux changes to match the difference between glucose uptake and oxidative disposal.

Elevated plasma lactate might explain the failure of epinephrine to stimulate glycogenolysis in the studies described in Section 7.3.2.3. To test this, lactate or pyruvate were infused to raise the circulating levels of these nutrients prior to a 10 min epinephrine infusion. The epinephrine infusion did not stimulate a further increase in the rate of glycogen synthesis when it was already stimulated by elevated lactate or pyruvate. This result, along with the studies described above, shows that the standard picture of heart glycogen synthesis largely being regulated by hormones is incomplete. The substrate environment of the heart can have as strong or even a greater influence on glycogen synthesis than the hormonal milieu, and cannot be neglected, even at physiological levels of lactate and ketones.

7.4. EFFECTS OF DIABETES ON HEART GLYCOGEN METABOLISM

Diabetes is characterized by elevated plasma glucose caused either by deficient insulin secondary to pancreatic beta cell injury, or by tissue insulin resistance. It is associated with both coronary artery disease due to increased atherosclerosis, and a specific diabetic cardiomyopathy. Cardiac dysfunction is observed in rat and mouse models of both type 1 (insulin-dependent) and type 2 (non-insulin-dependent) diabetes. All are characterized by reduced heart function, reduced glucose uptake and metabolism, elevated plasma triglycerides, fatty acids and beta oxidation, and, paradoxically, by elevated myocardial glycogen despite low insulin-stimulated GSase activity. Interestingly, prolonged fasting is also accompanied by low insulin, reduced glucose metabolism, elevated plasma fatty acids and triglycerides, and, similar to diabetes, elevated heart glycogen (21).

^{13}C NMR was used in order to test whether the increased glycogen concentration could be accounted for by the acute rates of glycogen synthesis and whether the changes in glycogen metabolism in a low-insulin diabetes model were similar to those observed during prolonged fasting (21). Rats were made diabetic by injection of 60 mg/kg body weight of alloxan 3 days before the experiment, or fasted for 24–65 h before the experiment. When measured in rats infused with infused insulin and [1-^{13}C]glucose, the rate of synthesis of myocardial glycogen was depressed in diabetes (0.18 ± 0.04 μmol/min g wet weight) and fasting (0.16 ± 0.03) relative to healthy fed control rats (0.32 ± 0.04). Extracted GSase enzyme activity was increased by insulin in all states, but was depressed in diabetes (0.10 μmol/min g wet weight) and fasting (0.14) relative to controls (0.33).

The studies described in Section 7.3 on the regulation of glycogen synthesis by mitochondrial substrates may provide a partial explanation for the elevated glycogen found in diabetes models. Owing to the increased levels of oxidizable substrates like ketones and fatty acids which inhibit glycolysis, it is possible that glucose uptake, although kept low by the absence of insulin, is still greater than the rates of glycolysis and glucose oxidation. This would allow glycogen stores to be built up over time, however slowly. Consistent with this explanation is the similar increase in glycogen level and decrease in glycogen synthesis rate found in fasting, which is also accompanied by low insulin and elevated plasma ketones and lipids, but not by the associated hyperglycemia of diabetes.

The reduction of the response to insulin in rats exposed to chronic diabetes and fasting may, in analogy with muscle, be related to the sustained increase in fatty acid levels. As described in Chapter 9, sustained plasma levels of free fatty acids lead to the down-regulation of the branch of the insulin signaling pathway which regulates glycogen metabolism. If the enzymes of the heart glycogen synthesis pathway undergo similar proportional up- and down-regulation as in muscle, then a similar impact on insulin-sensitive glycogen synthesis would be expected in heart. Further evidence for this hypothesis is the finding of lipid accumulation in hearts of diabetic animals, similar to that found in skeletal muscle made insulin-resistant by extended exposure to elevated fatty acids.

7.5. A METABOLIC CONTROL MODEL OF HEART GLYCOGEN SYNTHESIS

The continuing questions raised by the heart's ability to modulate glycogen metabolism in the presence of lactate, fatty acids or ketone bodies can be answered by MCA of the roles these metabolites play in the bifurcated pathway of glucose metabolism. As shown above, ^{13}C NMR has been used to study heart glycogen metabolism under a diverse range of hormonal and substrate conditions. The response of heart glycogen metabolism to these different conditions is complex. However we believe that they can be explained by analogy to what has been learned from studies of skeletal muscle glycogen synthesis. In this section we propose an MCA model based on that of muscle found in Chapters 3–5, with modifications based upon the greater relative glycolytic flux in the insulinized heart than in the resting muscle. We then show that this model is supported by the results described above from studies of hormones and substrates and by recent data from transgenic mice. Although at present only a proposal, this model can be quantitatively tested using experimentation methods similar to those used in muscle and other systems (Chapters 3 and 5). Quantitative understanding of the control of heart glycogen metabolism should be highly valuable for interpreting the cause of the changes in glycogen metabolism observed in diabetes and other diseases.

7.5.1. The MCA Model

In the heart as in muscle, glucose transport/phosphorylation has a high flux control coefficient for insulin-stimulated glycogen synthesis, while GSase has a low flux control coefficient. Insulin causes a proportional activation of glucose transport, hexokinase and GSase and elevated G6P levels. The low flux control exerted by GSase under these conditions is probably aided by its high elasticity to G6P concentration (sensitivity to allosteric activation), whereas glucose transport and hexokinase flux are less sensitive to G6P concentration when activated by insulin, and therefore exert more flux control over the pathway. The glycolytic pathway also has a high flux control coefficient for glycogen synthesis, and the heart model thereby differs from that of skeletal muscle. This is a consequence of the fact that glycolysis is partly regulated by nutrients that enter the TCA cycle downstream of the glycogen branch. Therefore, under insulin-stimulated conditions the rate of glycogen synthesis is largely determined by the difference between the rates of glucose transport/phosphorylation and glycolysis.

The G6P levels found in the insulinized heart are permissively high, which activates GSase and allows the flux through this enzyme into glycogen to equal this difference without large changes in G6P. Therefore, we postulate that, in analogy to muscle, G6P levels in the insulinized heart are relatively stable regardless of the partitioning of flux between glycogenesis and glycolysis.

7.5.2. Enhancement of Glycogen Synthesis by Insulin, Evidence for a High Flux Control Coefficient of Glucose Transport/Hexokinase

GLUT-4 is the major insulin-sensitive glucose transporter in heart. This protein is stored in cytoplasmic vesicles, which translocate to the plasma membrane when activated by insulin, ischemia and hypoxia. The increase in glycogen synthesis flux in response to an increase in insulin-stimulated GLUT-4 activity is direct support for a high control coefficient for glucose transport (or in combination with hexokinase which also may be insulin-stimulated). A limitation of this argument is that insulin also stimulates GSase. However, when glucose transport is constitutively activated in transgenic mice, basal glycogen synthesis flux is also increased in the absence of elevated insulin. As shown in Section 7.5.5, these recent studies support the concept that glucose transport/hexokinase maintains a high control coefficient for glycogen deposition.

7.5.3. Enhancement of Glycogen Synthesis by Plasma Lactate: Evidence for a High Flux Control Coefficient of Glycolysis

As described above, elevated plasma lactate leads to a several-fold increase in insulin-stimulated glycogen synthesis. This result can be explained by lactate acting, through several potential mechanisms (see

Section 7.5.4), to inhibit glycolysis without significantly reducing the insulin-stimulated glucose transport and phosphorylation (13). It is well documented that additional oxidizable substrates only slightly reduce glucose uptake and phosphorylation in the presence of insulin (13, 22). When glycolytic flux is reduced to a rate below that of glucose transport/phosphorylation, G6P concentration could rise, which would stimulate glycogen synthesis. [G6P] measured in glucose-perfused working rat hearts is elevated when acetoacetate or lactate are added to the perfusate, which can explain most of the 2-fold increase in glycogen synthesis measured (23, 24). However, this was observed in the insulin-free state. Insulin itself causes G6P levels in perfused heart to go up dramatically, but they are not then further increased with the addition of acetoacetate (24) or high fat, lactate, acetate or hydroxybutyrate (17, 22). This implies that, as in skeletal muscle, G6P levels are maintained more or less constant in insulinized heart by the competition between GSase and glycolytic fluxes.

7.5.4. Molecular Mechanism Coupling Alternate Substrate Utilization, Glycolysis and Glycogen

The inhibition of glycolysis, and shunting of glucose carbons to glycogen, induced by elevated lactate levels leaves open the question of what are the molecular mechanisms which lead to glycolytic inhibition? Since all substrates that are readily oxidized produce the same phenomenon, the energy state of the mitochondria may be the ultimate sensor. In fact, all substrates tested that result in increased glycogen synthesis from glucose also cause an increase in the mitochondrial redox potential and the cytosolic phosphorylation potential in perfused hearts (25) and in increased PCr/ADP *in vivo* (26). This can be communicated into the regulation of the glycolytic pathway in a number of ways. First, the reduced mitochondrial redox state may produce a reduced cytosolic redox state via pressure on the malate aspartate shuttle, resulting in inhibition of GAPDH. Evidence that this is not a primary mechanism comes from the fact that lactate, which increases NADH in the cytosol, and pyruvate, which reduces cytosolic NADH, equally upregulate glycogen production. The phosphorylation potential and ATP are similar with the two substrates, which together with mitochondrial NADH would inhibit pyruvate dehydrogenase. However, the lack of excess lactate produced from glucose argues that glycolysis is inhibited further up the pathway. Citrate concentration appears to be a likely candidate. Citrate is made from excess TCA cycle intermediates, and inhibits PFK. Citrate is increased 5-fold in working glucose-perfused rat heart when acetoacetate is provided (24) and between 2- and 10-fold in isolated cardiomyocytes when lactate or pyruvate is added to glucose-containing perfusate (22). Citrate made from labeled lactate or pyruvate is elevated in ^{13}C NMR spectra of the *in vivo* dog heart, and appears to have approximately the same concentration as glutamate under these conditions (26).

Finally, if glycolysis were not inhibited, and excess G6P were diverted to lactate and spilled from the cell when other oxidizable substrates were presented to the heart, there would be less substrate for glycogen synthesis. This doesn't appear to happen, as [3-^{13}C]lactate is not observed arising from labeled [1-^{13}C]glucose in hearts of intact dogs when other, unlabeled substrates are present at elevated concentrations (13). Although glucose-perfused hearts produce lactate, this doesn't appear to happen in normal aerobic heart, in intact animals, where lactate is a preferred substrate, and lactate balance is almost always positive. In fact, when labeled [3-^{13}C]pyruvate is infused into the LAD in dogs, lactate extraction is suppressed but very little [3-^{13}C]lactate (produced within the cardiac tissue) is seen in the *in vivo* ^{13}C-NMR spectrum (26). These observations suggest that, in intact animals, normal plasma concentrations of lactate are sufficient to maintain a state of lactate uptake, and lactate is not released by the heart even under conditions where insulin is present and there is an abundant supply of glucose and other oxidizable substrates. Therefore, in normal aerobic heart, the rates of glycolysis and pyruvate entry into the TCA cycle are similar.

7.5.5. Studies of Glycogen Metabolism in Transgenic Mice: Evidence for Control at the Glucose Transport/Phosphorylation Steps

A prediction of the MCA model is that an increase in the activity of glucose transport or phosphorylation will lead to an increase in glycogen synthesis. Several transgenic mouse models have recently been produced that support this prediction, including myocardial hexokinase and GLUT-4 insulin-stimulated glucose transporter transgenic mice, and mice lacking GLUT-4. Mouse hearts that overexpress yeast hexokinase have constitutively elevated rates of glucose phosphorylation and glycolysis and, although GSase activity is only slightly activated, the concentration of glycogen in these hearts is more than double that found in control animals. G6P, on the other hand, was similar in both hexokinase transgenics and controls in the basal state, and it was similarly elevated by more than 2-fold in both groups with insulin (27). When GLUT-4 was overexpressed by 40 % in heart, this resulted in a more than 2-fold elevation of basal glucose uptake and glycolysis. Again, glycogen concentration was appreciably higher in the transgenics (3.9 vs 2.8 mg/g wet weight). Sixty minutes of perfusion with insulin activated glucose uptake to the same level in transgenics and controls, and this resulted in similarly elevated glycogen levels (5.6 vs 5.7 mg/g wet weight) (28). The db/db leptin receptor-deficient mouse is a model of type 2 diabetes, It is highly insulin-resistant and shows reduced insulin-stimulated cardiac glucose transport and glycogen synthesis. An increase in glucose transporter activity by overexpression of human GLUT-4 protein resulted in increased glucose uptake and oxidation, as well as increased glycogen storage (28, 29). GLUT-4 overexpression also acted to restore cardiac function, highlighting the important role of cardiac glucose and glycogen metabolism. Taken together, all these models strongly support the idea that glucose uptake and phosphorylation exert high control strength over glycogen synthesis in the heart.

In contrast to these transgenic models, it could be predicted that insulin-stimulated glucose and glycogen metabolism would be decreased in a GLUT-4 knockout mouse, if glucose transport is important in control of these pathways. ^{13}C (27) and ^{31}P NMR (30–32) were used to study glucose and glycogen metabolism of Langendorf perfused hearts from these mice, where the perfusate included ^{13}C-labeled glucose, palmitate, β-hydroxybutyrate and insulin. Heart function in fed GLUT-4 null mice was similar to control, but PCr/ATP was increased by 60 % in fed GLUT-4 null hearts (31, 32). Interrogation of glucose and glycogen metabolism in the GLUT-4 knockout model yielded even more surprising results. ^{31}P NMR was used to measure glucose phosphorylation as 2-deoxyglucose-6-phosphate accumulation, which was found to be increased two-fold in GLUT-4 deficient hearts. This rate of accumulation was unresponsive to insulin. The [1-^{13}C]glycogen synthesis rate was increased 3-fold over controls in GLUT-4 deficient hearts (30).

What could explain the increased glucose uptake and glycogen synthesis in this model? The most likely explanation is that these animals develop a compensating glucose transporting mechanism, and it was subsequently shown that they have a 3-fold elevated myocardial expression of the insulin-insensitive GLUT-1 (33). Thus, the GLUT-4 knockout mouse is a model of constitutively active glucose transport, where glycogen storage and synthesis rates are elevated. Once again, this provides strong evidence that glucose uptake exercises a high degree of flux control in the pathway of glycogen synthesis. It is of great interest that the genetic manipulated animals, designed to answer a question about the role of GLUT-4, could only be useful after metabolic studies had shown that the expectations from a purely genetic standpoint were incorrect.

ABBREVIATIONS

tca	tricarboxylic acid cycle, Kreb's cycle, citric acid cycle
gsase	glycogen synthase
gp	glycogen phosphorylase

PCr	phosphocreatine
udpg	uridine diphosphoglucose
G1P	glucose-1-phosphate
g6p	glucose-6-phosphate
glut-4	glucose transporter-4

REFERENCES

1. Roach PJ. Glycogen and its metabolism. *Curr Mol Med.* 2002; **2**(2): 101–120.
2. Jacobus WE, Taylor GT, Hollis DP, Nunally RI. Phosphorus nuclear magnetic resonance of perfused working rat hearts. *Nature* 1977; **265**: 756–758.
3. Grove TH, Ackerman JJH, Radda GK, Bore PJ. Analysis of heart *in vivo* by phosphorus nuclear magnetic resonance, *Proc Natl Acad Sci USA* 1980; **77**: 299–302.
4. Neurohr KJ, Barrett EJ, Shulman RG. *In vivo* carbon-13 nuclear magnetic resonance studies of heart metabolism. *Proc Natl Acad Sci USA* 1983; **80**(6): 1603–1607.
5. Neurohr KJ, Gollin G, Neurohr JM, Rothman DL, Shulman RG. Carbon-13 nuclear magnetic resonance studies of myocardial glycogen metabolism in live guinea pigs. *Biochemistry* 1984; **23**(21): 5029–5035.
6. Zang LH, Laughlin MR, Rothman DL, Shulman RG. ^{13}C NMR relaxation times of hepatic glycogen *in vitro* and *in vivo*. *Biochemistry* 1990; **29**(29): 6815–6820.
7. Botker HE, Randsbaek F, Hansen SB, Thomassen A, Nielsen TT. Superiority of acid extractable glycogen for detection of metabolic changes during myocardial ischaemia. *J Mol Cell Cardiol* 1995; **27**(6): 1325–1332.
8. Sillerud LO, Shulman RG. Structure and metabolism of mammalian liver glycogen monitored by carbon-13 nuclear magnetic resonance. *Biochemistry* 1983; **22**(5): 1087–1094.
9. Garlick PB, Pritchard RD. Absolute quantification and NMR visibility of glycogen in the isolated, perfused rat heart using ^{13}C NMR spectroscopy. *NMR Biomed* 1993; **6**(1): 84–88.
10. Hoekenga DE, Brainard JR, Hutson JY. Rates of glycolysis and glycogenolysis during ischemia in glucose–insulin–potassium-treated perfused hearts: a ^{13}C, ^{31}P nuclear magnetic resonance study. *Circul Res* 1988; **62**(6): 1065–1074.
11. Laughlin MR, Petit WA Jr, Dizon JM, Shulman RG, Barrett EJ. NMR measurements of *in vivo* myocardial glycogen metabolism. *J Biol Chem* 1988; **263**(5): 2285–2291.
12. Laughlin MR, Taylor JF, Chesnick AS, Balaban RS. Regulation of glycogen metabolism in canine myocardium: effects of insulin and epinephrine *in vivo*. *Am J Physiol* 1992; **262**(6 Pt 1): E875–883.
13. Laughlin MR, Taylor J, Chesnick AS, Balaban RS. Nonglucose substrates increase glycogen synthesis *in vivo* in dog heart. *Am J Physiol* 1994; **267**(1 Pt 2): H219–223.
14. Brainard JR, Hutson JY, Hoekenga DE, Lenhoff R. Ordered synthesis and mobilization of glycogen in the perfused heart. *Biochemistry* 1989; **28**(25): 9766–9772.
15. Laughlin MR, Petit WA Jr, Barrett EJ. The time course of myocardial glycogenolysis stimulated by glucagon. *J Mol Cell Cardiol* 1993; **25**(2): 175–183.
16. Goodwin GW, Ahmad F, Doenst T, Taegtmeyer H. Energy provision from glycogen, glucose, and fatty acids on adrenergic stimulation of isolated working rat hearts. *Am J Physiol* 1998; **274**(4 Pt 2): H1239–1247.
17. Goodwin GW, Taegtmeyer H. Improved energy homeostasis of the heart in the metabolic state of exercise. *Am J Physiol Heart Circul Physiol* 2000; **279**(4): H1490–1501.
18. Laughlin MR, Morgan C, Barrett EJ. Hypoxemic stimulation of heart glycogen synthase and synthesis. Effects of insulin and diabetes mellitus. *Diabetes* 1991; **40**(3): 385–390.
19. Sambandam N, Lopaschuk GD, Brownsey RW, Allard MF. Energy metabolism in the hypertrophied heart. *Heart Fail Rev* 2002; **7**(2): 161–173.
20. Laine H, Yki-Jarvinen H. Kirvela O, Tolranen T, Raitakari M, Solin O, Haaparanta M, Knuuti J, Nuutila P. Insulin resistance of glucose uptake in skeletal muscle cannot be ameliorated by enhancing endothelium-dependent blood flow in obesity. *J Clin Invest* 1998; **101**(5): 1156–1162.

21. Laughlin MR, Petit WA Jr, Shulman RG, Barrett EJ. Measurement of myocardial glycogen synthesis in diabetic and fasted rats. *Am J Physiol* 1990; **258**(1 Pt 1): E184–190.
22. Fischer Y, Bottcher U, Eblenkamp M, Thomas J, Jungling E, Rosen P, Kammermeier H. Glucose transport and glucose transporter GLUT4 are regulated by product(s) of intermediary metabolism in cardiomyocytes. *Biochem J* 1997; **321**: 629–638.
23. Russell R, Cline GW, Guthrie PH, Goodwin GW, Shulman GI, Taegtmeyer H. Regulation of exogenous and endogenous glucose metabolism by insulin and acetoacetate in the isolated working rat heart. *J Clin Invest* 1997; **100**: 2892–2899.
24. Depre C, Veitch K, Hue L. Role of fructose 2,6-bisphosphate in the control of glycolysis. Stimulation of glycogen synthesis by lactate in the isolated working rat heart. *Acta Cardiol* 1993; **48**(1): 147–164.
25. Laughlin MR and Heineman FW. The relationship between phosphorylation potential and redox state in the isolated working rabbit heart. *J Mol Cell Cardiol* 1994; **26**(12): 1525–1536.
26. Laughlin MR, Taylor J, Chesnick AS, DeGroot M, Balaban RS. Pyruvate and lactate metabolism in the *in vivo* dog heart. *Am J Physiol* 1993; **264**: H2068–H2079.
27. Liang Q, Donthi RV, Kralik PM, Epstein PN. Elevated hexokinase increases cardiac glycolysis in transgenic mice. *Cardiovasc Res* 2002; **53**: 423–430.
28. Belke DD, Larsen TS, Gibbs EM, Severson DL. Glucose metabolism in perfused mouse hearts overexpressing human GLUT-4 glucose transporter. *Am J Physiol Endocrinol Metab* 2001; **280**(3): E420–427.
29. Belke DD, Larsen TS, Gibbs EM, Severson DL. Altered metabolism causes cardiac dysfunction in perfused hearts from diabetic (db/db) mice. *Am J Physiol Endocrinol Metab* 2000; **279**(5): E1104–1113.
30. Stenbit AE, Katz EB, Chatham JC, Geenen DL, Factor SM, Weiss RG, Tsao TS, Malhotra A, Chacko VP, Ocampo C, Jelicks LA, Charron MJ. Preservation of glucose metabolism in hypertrophic GLUT4-null hearts. *Am J Physiol Heart Circul Physiol* 2000; **279**(1): H313–318.
31. Weiss RG, Chatham JC, Georgakopolous D, Charron MJ, Wallimann T, Kay L, Walzel B, Wang Y, Kass DA, Gerstenblith G, Chacko VP. An increase in the myocardial PCr/ATP ratio in GLUT4 null mice. *FASEB J* 2002; **16**(6): 613–615.
32. Tian R, Abel ED. Responses of GLUT4-deficient hearts to ischemia underscore the importance of glycolysis. *Circulation* 2001; **103**(24): 2961–2966.
33. Abel ED, Kaulbach HC, Tian R, Hopkins JC, Duffy J, Doetschman T, Minemann T, Boers ME, Hadro E. Cardiac hypertrophy with preserved contractile function after selective deletion of GLUT4 from the heart. *J Clin Invest* 1999; **104**: 1703–1714.

8

Bioenergetics Implication of Metabolic Fluctuation during Muscle Contraction

Thomas Jue

Department of Biochemistry and Molecular Medicine, University of California Davis, Davis CA 95616-8635, USA

Metabolomics by In Vivo NMR. Edited by R. G. Shulman and D. L. Rothman
 ISBN: 0-470-84719-0

8.1. HISTORICAL PERSPECTIVE

8.1.1. Introduction

Over the years, physiologists have established a paradigm to link the transduction of mechanical, thermal and chemical energy during muscle contraction. Based primarily on steady-state measurements, the conventional view downplays any significant role for metabolic transients. Yet many experimental observations appear incongruent with such a strict exclusion of metabolic fluctuation. With the advent of *in vivo* NMR, techniques became available to measure reaction kinetics in the cell. Chung *et al.* (1998) developed a gated NMR technique to observe the transient fluctuation in phosphocreatine (PCr) during a contraction cycle and noted that PCr decreased dramatically during a twitch. Their findings have challenged the current bioenergetics paradigm in muscle contraction and have stimulated a re-examination about how the cell regulates the transient and steady-state energy flow. To participate meaningfully in that discussion requires a historical perspective of the muscle bioenergetics, an identification of the various underlying assumptions in the interpretation of experimental data, a synopsis of the new experimental data, and perspectives on the potential implications.

8.1.2. Heat and Work

Muscle contraction originates from filament movement, and the energy powering the filament motion depends upon a transduction of chemical energy into mechanical work and therefore an interplay between heat and work. The energy conversion must then obey the first law of thermodynamics, $\Delta E = q + w$, where ΔE = energy, q = heat and w = work. Hill and his colleagues leaned on this thermodynamic approach to probe the relationship between heat production and force development or work (Hill, 1965). If work originates only from mechanical displacement and excludes all other work, including any electrical work contribution associated with ion movement, then during the specific condition of isometric contraction $w = 0$, since no cross-bridge movement presumably occurs. Heat, or q, then equals ΔE.

Hill focused his analysis on the heat and velocity of shortening to clarify a model of muscle contraction (Aidley, 1989). At that time physiologists viewed muscle responding to nerve stimulation as a stretched elastic body. An elastic body, however, dissipates its potential energy to generate a final force determined only by the stretch length. The force should not depend upon any velocity of shortening. Yet, investigators noted that the force of isotonic shortening did depend on the velocity, leading them to postulate the presence of an internal viscosity. The muscle must work against this internal viscosity, so that tension develops more quickly at high velocity during the onset of contraction, and that the tension development originates primarily from a viscous region not at the ends of the muscle. This supposition about an elastic body and viscous damping formed the basis of the viscoelastic theory of muscle contraction. Indeed, until 1938, the viscoelastic theory held sway, even though Fenn in 1923 had presented evidence to question its validity. He had already noted that the muscle released more energy when it could shorten, which did not conform to the expected response of an elastic body (Fenn, 1923, 1924; Levin and Wyman, 1927).

Hill's analysis assumed only displacement work and related the additional energy liberated during shortening to the heat of shortening and mechanical work:

$$\Delta E_a = q_s + w_m = a \cdot x + P \cdot x$$

where P = tension created by different loads, a = constant related to heat of shortening, and x = displacement. Then

$$dE_a/dt = [(P + a)x]/dt = (P + a)V$$

where V = velocity.

Since thermal measurement can follow the heat of shortening, experiments can then determine how dE_a/dt varies with P and V. In fact, dE_a/dt decreased in proportion to increasing tension,

$$dE_a/dt = b(P_o - P).$$

These two equations lead to

$$(P + a)V = b(P_o - P) \text{ or } (P + a)(V + b) = b(P_o + a) = \text{constant}$$

where P_o is the isometric tension when no shortening occurs.

Substituting the experimentally determined values of a, b and P_o, the hyperbolic function of the last equation provides an excellent curve fit of the velocity of shortening vs load or the force velocity profile (Hill, 1938).

Hill then created a model to account for the mathematical behavior. The model contained an elastic component in series with the contractile component. Later the model incorporated a parallel elastic component. At the onset of contraction, the contractile component begins to shorten. Since the resting state tension is zero, the initiation of contraction starts at maximum velocity. However, the shortening of the contractile component produces a corresponding extension of the elastic component, creating a rise in tension. As tension rises, the velocity of shortening slows down, which in turn reduces any additional increase in tension development. Eventually, shortening velocity will decrease to zero and the elastic component becomes fully extended (Hill, 1938). The model then explained the interaction of force and velocity during contraction, as reflected in the force velocity profile.

Physiologists have built on Hill's findings to compose a molecular view of muscle tissue, comprising blood perfused muscle fiber bundles (Aidley, 1989; McComas, 1996). Each fiber in the bundle contains many parallel myofibrils, typically 1 μM in diameter. In turn, each myofibril has a number of repeating functional units or sacromeres distributed every 2.3 μm along the longitudinal axis. The schematic in Figure 8.1 depicts a typical sacromere. As muscle contracts, neither the thick nor thin filament changes its length. Instead they slide across each other and pull the Z lines closer to contract the overall sacromere unit length. These observations form then the basis of the sliding filament model of muscle contraction, first proposed in the 1954 (Huxley and Niedergerke, 1954; Huxley and Hanson, 1954).

Although Hill's series elastic model has a number of limitations, it has spurred the search to identify the molecular component associated with the elasticity. Researchers have ascribed the elastic component to the muscle tendons and the cross bridges (Huxley and Simmons, 1971, 1973). Ironically though the presence of the elastic component in the cross bridges argues against the accuracy of Hill's two-component model (Aidley, 1989).

However, Hill recognized he had only cast the first light on the mechanism of muscle contraction, so in 1950 he openly exhorted biochemists to search for the immediate source of chemical energy required for contraction. What actually fuels filament movement (Hill, 1950)?

8.1.3. Energy Source

The filament sliding movement must harness chemical energy to generate the force that contracts the sacromere length. Initially, the search for the contraction energy source focused on the glycolytic pathway

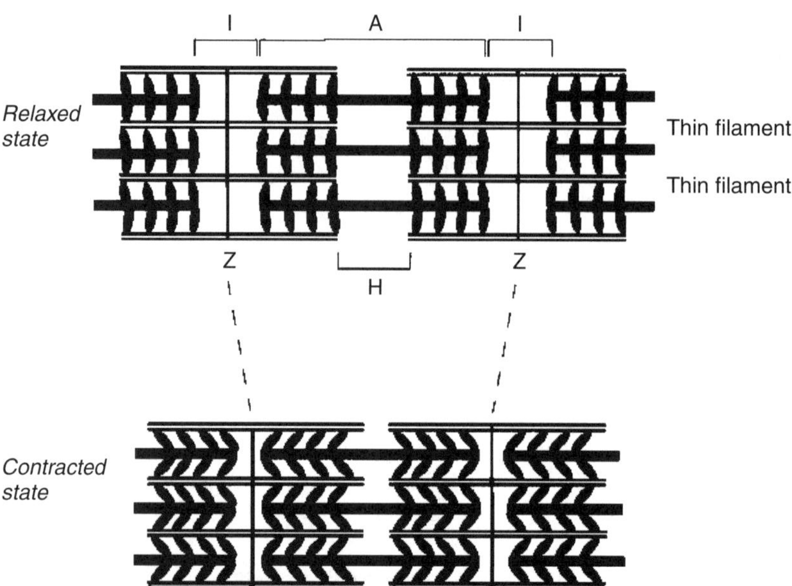

Figure 8.1. Diagram of the sarcomere. Muscle is composed of fibers. Each fiber in the bundle contains many parallel myofibrils, typically 1 μM in diameter. In turn, each myofibril has a number of repeating functional units or sacromeres distributed every 2.3 μm along the longitudinal axis. Under a light or phase contrast microscope, each sacromere exhibits a distinct striation pattern, alternating dark (A) and light (I) bands, an H zone, an I zone and a Z line. These features reflect the spatial distribution of two kinds of protein filaments: the thick filament (15 nm in diameter) and the thin filament (9 nm in diameter). Overlapping thick and thin filaments produce the dark A bands punctuated in the center with a light H zone, which contains only thick filaments. In contrast, leading away from the central A band, the I zone with only thin filament terminates at the sacromere boundary or Z line. The presence of myosin and actin characterizes the thin from the thick filaments. Myosin (520 kDa) constitutes the thick filament and is composed of six polypeptide chains: two identical heavy chains (200 kDa) and two sets of light chains (20 kDa each). The heavy chains form both a rod-like tail and a globular head structure. In the globular region, the two different light chains bind: essential light chain (ELC) and regulatory light chain (RLC). Proteolysis cleaves myosin into two functional fragments: light meromysosin (LMM) and heavy meromyosin (HMM). The LMM fragment forms the rod-like tail region, while the HMM fragment, comprising the S1 and S2 subfragments, forms another portion of the tail section but also the head region. At the globular head region, ATP and actin bind to specific sites. In contrast, actin is the major constituent of the thin filament, which contains also troponin and tropomyosin. Actin exists as a 42 kDa monomer (G actin) but under physiological conditions polymerizes and forms F-actin. (Illustration by Ulrike Kreutzer.)

(Figure 8.2). Fletcher and Hopkins (1907) noted that muscle could contract under anaerobic conditions and produced lactate. The observation led to the supposition that the reaction leading to lactate formation supplied the immediate source of energy and formed the basis of the 'lactic acid theory'. This idea held sway until Lundsgaard (1930a, b) measured the lactate and PCr concentration in contracting anaerobic muscle inhibited with iodoacetate. With glyceraldehydes 3 phosphate dehydrogenase (GAPDH) inhibited by iodoacetate, glycolysis could no longer produce lactate. Yet muscle could still contract with the expenditure of PCr. Others researchers had also noted that PCr level decreased in proportion to muscle tension (Eggleton and Eggleton, 1927; Fiske and Subbarrow, 1927). Since lactate acid did not serve as the immediate energy precursor for muscle contraction, the 'lactic acid theory' collapsed.

Subsequent analysis began to compose a different view: adenosine triphosphate (ATP) could provide the energy source, where the cleavage of the two terminal phosphate bonds yielded chemical energy to drive

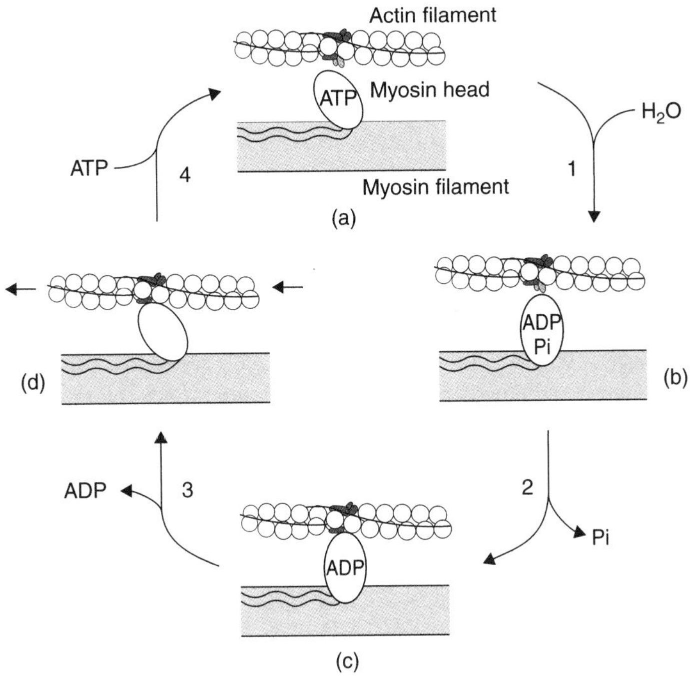

Figure 8.2. ATP reaction in myosin. In resting muscle, tropomyosin sterically hinders the S1 heads of myosin from interacting with actin in the thin filament. In this state, ADP and P_i remain bound to myosin. Upon nerve excitation, the sarcoplasmic reticulum releases Ca^{2+}, which binds to the troponin complex, specifically troponin C (TnC) and induces a conformational change in myosin. As a result, tropomyosin no longer hinders the binding of myosin to actin, and myosin releases the bound ADP and P_i. As the S1 head binds to actin it moves the myosin protein approximately 100 Å along the thin filament and effectively draws the Z lines closer together. ATP can now bind to myosin and induces another conformation that releases the binding to actin. Myosin then hydrolyzes ATP to ADP and P_i and returns to its resting state. A repeating cycle of ATP binding, ATP hydrolysis, ADP and P_i release; of S1 myosin binding to and releasing from actin; and of myosin ratcheting in a cross-bridge movement along the thin filament constitute the molecular steps that transduce chemical energy into mechanical force development during muscle contraction. (Illustration by Ulrike Kreutzer.)

cellular reactions. Both the creatine kinase and glycolysis pathways synthesized ATP. In glycolysis, the catabolism of glucose or glycogen to pyruvate produced two or three net ATP molecules, and pyruvate converted readily to lactate, especially under anaerobic conditions. PCr, however, could transfer its high phosphate potential group to adenosine diphosphate (ADP) to rapidly form ATP (Lipmann, 1941; Lohmann, 1934; Meyerhof and Lohmann, 1932; Needham, 1971).

With either glycolysis or creatine kinase capable of responding to the energy demand in muscle contraction, how did the cell order the energy withdrawal? Indeed, the experimental observations suggested that the PCr reaction, catalyzed by creatine kinase, a near-equilibrium enzyme, served as an energy buffer. Several lines of evidence supported the supposition: with iodoacetate inhibition of glycolysis, muscle could still contract, and the force corresponded proportionally to PCr breakdown. Moreover, complete glycolytic ATP synthesis required nine reaction steps, whereas PCr-catalyzed ATP formation needed only one reaction step. Since the creatine kinase-catalyzed reaction could operate much faster than glycolysis, and if contracting muscle actually utilized ATP to fuel contraction, PCr served as an ATP energy buffer, back-filled subsequently by glycogenolysis or glycolysis and other ATP-generating reactions.

Although experiments had correlated a decline of PCr during muscle contraction, they had not actually demonstrated that cross-bridge movement actually utilized ATP, drawn from PCr. Did a muscle contraction actually consume ATP?

Hill's challenge stimulated the development of rapid tissue freezing techniques to trap rapidly cellular metabolites during a twitch (Hill, 1950; Mommaerts, 1954). The early techniques that relied on immersing tissue in liquid nitrogen met with only modest success. Unfortunately, the gaseous insulating layer formed in liquid nitrogen obfuscated the freezing time resolution, but new methods began to emerge, culminating in Kretzschmar's apparatus, which used hammers cooled to −196 °C to snap-freeze within 100 ms the center of a piece of tissue to −10 °C (Kretzschmar and Wilkie, 1969). With the new techniques, the ensuing experiments confirmed that PCr did decline during a muscle contraction and could buffer ATP (Infante and Davies, 1962; Infante *et al.*, 1965). Cain *et al.* (1962a, b) then introduced 1 fluoro-2,4 dinitrobenzene to block the creatine kinase activity. They could then observe ATP breakdown during a twitch. Seven years after Hill issued his challenge, biochemists presented experimental evidence to support the theory that PCr acted as an energy buffer, and ATP served as the immediate energy source during a muscle contraction

8.1.4. Energy Balance

Cain *et al.* (1962a, b) achieved a milestone in biochemistry by studying a model system uncoupled from the normal physiological controls. Only with frog rectus abdominis muscle poisoned with a glycolysis inhibitor and a respiration uncoupler could the investigators determine the balance between ATP hydrolysis energy and heat plus work. Using these inhibitors, Cain *et al.* overcame key experimental obstacles: because normal frog rectus abdominis muscle exhibited a relatively low concentration of inorganic phosphate (P_i) and creatine (Cr; about 2.5 and 5.0 μmol/g), freeze-trapping and subsequent biochemical analysis could discriminate accurately these metabolite levels before and after a contraction. However, the high PCr level still posed a hurdle, since any small change would avoid detection, given the dynamic fractional change in the metabolite pool. To improve the detection sensitivity, the experimenters introduced a respiratory chain uncoupler 2,4 dinitrophenol (DNP) to reduce the efficiency of aerobic ATP production. The control PCr level then dropped from 12 to 2.5 μmol/g. Finally, 1-fluoro-2,4 dinitrobenzene then inhibited the creatine kinase activity for the ATP transient measurements. At peak contraction with surrounding temperature at 0 °C, the experimental apparatus quickly immersed the muscle into a 1:1 mixture of dichlorodifluoromethane (CF_2Cl_2) and fluorotrichloromethane ($CFCl_3$) at −172 °C. The separation of P_i, PCr and Cr concentration from the frozen tissue employed aqueous methanol extraction at −35 °C for 7 days, following Wahler and Wollenberger's (1958) quantitation method. Despite the unphysiological conditions, the experiments captured the twitch cost estimate of 0.36 μmol of ATP per gram tissue (Infante *et al.*, 1965). Given the ATP hydrolysis energy of 47 kJ/mol, ATP supplied approximately 10^{-5} kJ/mol/g tissue/twitch.

Wilkie studied frog sartorius muscle to map the relationship between chemical energy released in ΔPCr and the energy manifested in heat + work during isometric twitch, tetanus and isotonic tetanus. He poisoned the tissue with iodoacetate to inhibit glycolysis and introduced nitrogen to block cellular respiration. Under these anaerobic, glycolysis-inhibiting conditions, PCr must phosphorylate ADP to ATP. In essence, any ΔPCr would reflect the net ATP chemical energy used during a contraction. Wilkie stimulated the muscle to contract between four and 107 times in order to vary the heat + work and then measured the corresponding ΔPCr. The analysis yielded a linear relationship with a slope of 46.4 kJ/mol, in excellent agreement with the free energy of ATP hydrolysis and with the notion that ATP provided all the energy required during a muscle contraction (Wilkie, 1968).

A subsequent re-evaluation of the expected enthalpy PCr hydrolysis lowered the estimate on the expected enthalpy ATP hydrolysis from 47 to 34 kJ/mol (Woledge, 1973). The chemical energy in ATP hydrolysis could no longer supply all the energy observed in the heat + work, and the discrepancy launched the quest to determine the source of the 'unexplained enthalpy' unassociated with cross bridge movement (Homsher *et al.*, 1981; Homsher and Kean, 1982; Woledge *et al.*, 1985). In contrast, the NMR-determined free ADP pool has led to a higher value for ATP hydrolysis. In cat bicep and soleus muscle, studies have reported 72 and 58 kJ/mole, respectively (Meyer *et al.*, 1985).

8.1.5. Bioenergetics Paradigm

Based on the evidence from these historical experiments, the current paradigm of muscle bioenergetics during a contraction has emerged (Kushmerick, 1983; Meyer and Foley, 1996): the initiation of muscle contraction starts with an action potential that travels along the surface plasmalemma and down the transverse tubules (T-tubules), where it stimulates the release of Ca^{2+} from the sarcoplasmic reticulum (Gage and Eisenberg, 1969; Huxley and Taylor, 1958). Mediated by the Ca^{2+} binding to troponin C, myosin hydrolyzes ATP, which then fuels cross bridge movement (Figure 8.3).

The cell responds to the increased energy or ATP demand by reacting PCr with ADP to form ATP and Cr via the creatine kinase-catalyzed reaction. PCr breakdown provides ATP energy for muscle contraction. During the rest interval in a contraction cycle, PCr level must regenerate. However, PCr does not restore

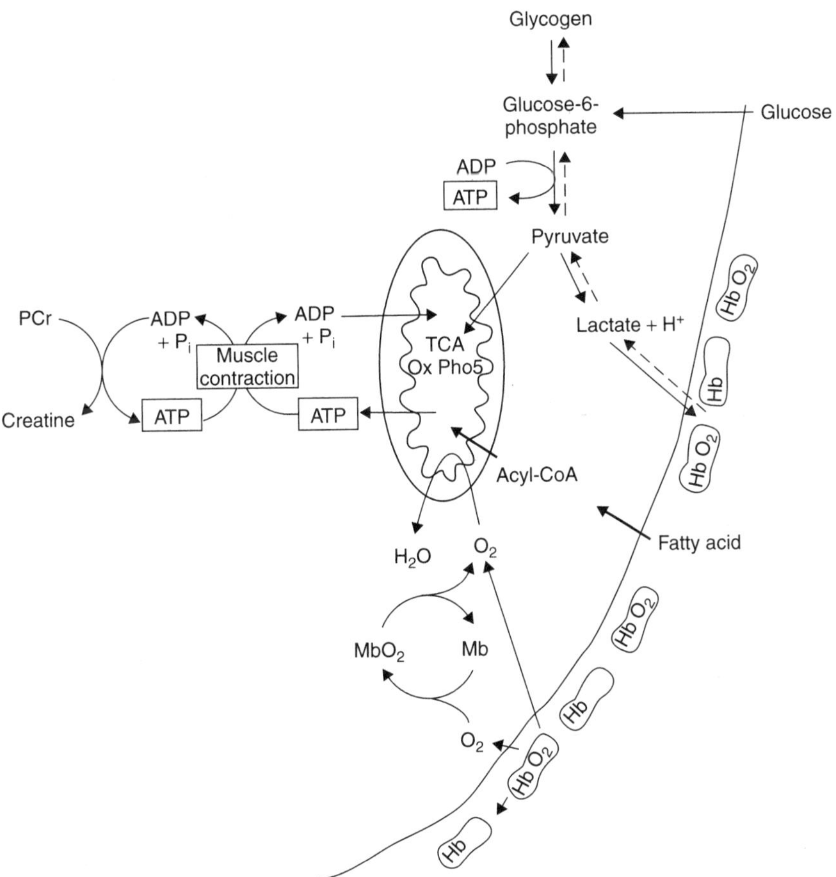

Figure 8.3. Cellular energy pathways. A schematic drawing of the cell and the major metabolic pathways that control energy utilization. (Illustration by Ulrike Kreutzer.)

rapidly from multi-step *de novo* synthesis. Instead, the replenishment of the ATP pool or the readjustment of the ATP/ADP ratio shifts the creatine kinase equilibrium toward PCr formation. From this vantage, PCr acts consistently as an ATP or energy buffer. Given such a view and the first law of thermodynamics, the chemical energy from PCr catabolism reflects ATP hydrolysis, which should equal all the released energy, including heat and work.

ATP generation or replenishment depends upon key metabolic pathways: glycolysis, glycogenolysis, and oxidative phosphorylation (Figure 8.2). These different interacting pathways regulate the rate of ATP metabolism and direct the bioenergetics toward a defined homeostasis. The control of bioenergetics, however, does not simply reside in the regulation of any particular metabolic step, for many interdependent metabolic steps intervene. For example, early investigators linked glycolysis as the immediate energy source to fuel muscle contraction. Although glycolysis produces ATP, which myosin uses for cross-bridge movement, it does not fuel the immediate energy need of muscle contraction. Instead, PCr does. The integrative metabolic response that directs the energy flow during muscle contraction still forms a central research perspective: how does the cell regulate its metabolic supply/demand to meet the energy needs during muscle contraction?

8.1.6. Advent of NMR

The advent of *in vivo* NMR has opened a unique opportunity to probe energy regulation of muscle contraction in normothermic tissue throughout a range of physiological conditions. Over the years, researchers have utilized ^{1}H, ^{13}C and ^{31}P NMR to interrogate the regulation of intermediary muscle metabolism in cells, isolated fibers, in whole animals, and in humans (Cozzone and Bendahan, 1994; Kushmerick and Meyer, 1985; McCully *et al.*, 1994; Shulman and Rothman, 2001a). In particular, the ^{31}P spectra readily reveal the signals of PCr, ATP, P_i and other notable metabolite signals, namely the phosphomonoesters (PME) and phosphodiester (PDE) (Dawson *et al.*, 1977; Radda, 1986). Unfortunately, the ^{31}P spectra do not yield a discernible ADP signal, required to determine the phosphorylation potential or energy charge of the cell. The spectral overlap of the ATP and ADP peaks and the low ADP cellular concentration preclude the direct detection of ADP.

To assess the ADP concentration, investigators have assumed creatine kinase at equilibrium and have used the observables in the ^{31}P spectra (PCr, ATP, P_i and pH) to calculate the ADP activity, which significantly exceeds the activity derived from biochemical assay determination. The difference in the NMR- vs biochemical assay-determined ADP concentration arises presumably from the distinction between the soluble, NMR-detectable ADP vs the total, bound and free, ADP in the cell (Iles *et al.*, 1985). Alternatively, Chance *et al.* (1981) have proposed the observable P_i/PCr ratio to index ADP. However, its validity presumes a stoichiometric relationship between P_i and Cr, a constant pH, and an unperturbed ATP (Chance and LaNoue, 1990; Connett and Honig, 1989, 1990).

With the detectable ^{31}P NMR signals, experiments show that during low-intensity muscle contraction the ATP in normal muscle remains quite constant, but PCr decreases exponentially (Taylor *et al.*, 1986). PCr declines more rapidly at high stimulation rate than at low rate, and the steady-state level also falls much lower. At high contraction rate, however, PCr no longer decays with a single exponential time constant.

8.1.7. Saturation Transfer

^{31}P NMR can certainly measure the steady-state level of high-energy phosphate metabolites and extract the pertinent signals to assess the energy charge of the cell, but it can also follow the dynamic steady-state flux in an enzyme-catalyzed reaction with magnetization transfer techniques. These techniques saturate or invert selectively either the reaction substrate or product and then monitor the fractional signal intensity loss or T_1 relaxation curve of the unperturbed signal as the saturated spin population exchanges between sites. The resultant analysis holds strictly for a two-site exchange and yields a unidirectional rate constant (Alger and Shulman, 1984).

Early magnetization saturation studies of the creatine kinase flux during rat skeletal muscle contraction between 0.25 and 2 Hz showed a constant PCr to ATP flux of about 16 mM/s, at least 10 times higher than the estimated ATP turnover derived from a presumed linear dependence of tension time integral and ATP turnover (0.1–0.5 mM/s) (Shoubridge *et al.*, 1984). Contrary to expectation, the increased ATP turnover with higher contraction frequency elicited no corresponding increase in the PCr to ATP flux. Moreover, the reverse flux from ATP to PCr appeared smaller than the forward flux from PCr to ATP, and the mismatched kinetics raised questions about the near-equilibrium assumption of the creatine kinase reaction and PCr compartmentalization in the cell (Matthews *et al.*, 1982; Meyer *et al.*, 1982, 1984; Nunnally and Hollis, 1979; Rees *et al.*, 1989). Some investigators then postulated that the NMR measurements might underestimate the MgADP concentration, which would obscure the determination of the PCr to ATP flux. The view captured a particular angst in the *in vivo* magnetic resonance community, especially in light of the reported mismatched in the creatine kinase forward and reverse fluxes, which would then preclude even the ADP calculation based on the critical near-equilibrium assumption.

These early saturation experiments, however, assumed a strict two-site exchange between PCr and ATP and accordingly saturated one site at a time. However, ATP can participate in more than one reaction pathway. So neither the model nor the associated experimental protocol had any rigorous validity. A multisite irradiation must saturate all ATP side reaction contributions in order to create an apparent two-site exchange model, amenable to magnetization transfer analysis. Indeed, such multisite saturation experiments confirmed that the PCr to ATP and ATP to PCr fluxes do match and that the assumption of a near-equilibrium creatine kinase had validity (Ugurbil, 1985).

8.2. TRANSITION VIEWS

8.2.1. Thermodynamics

Given such an extensive investigatory history of muscle bioenergetics, one might presume that no outstanding issues should still exist. Certainly all current experiments stand firmly on the scientific foundation erected by these early researchers. Yet a critical assessment of the historical evidence reveals that critical, pressing questions still remain. These questions rationalize the transition into a basis for undertaking current research.

Hill's analysis began with the first law of thermodynamics and equated the released energy with heat and work, q and w. The experiments measured directly heat and work and made the assumption that muscle acts as a heat machine, that all work arises from the mechanical work of cross-bridge movement, and that any change in the internal properties of muscle during contraction activity will not significantly alter the observation. The analysis, however, overlooked several key considerations: Fick had already presented thermodynamics arguments against the paradigm of muscle as a heat machine – the probability of a significant temperature gradient within muscle appeared low. As a heat machine, the thermodynamic efficiency would also be low. Consistent with the Carnot cycle analysis, decreasing temperature reduces the thermodynamic efficiency (Kushmerick, 1983). Moreover, the underlying assumptions preclude ion transients from impacting directly on either q or w. However, muscle contraction triggers significant ion movement through electric fields across membranes of the sarcoplasmic reticulum and mitochondria, producing significant electrical work. Hill's model analysis excluded any non-displacement work.

From the vantage of released energy, experiments have proven that contraction energy originates from ATP hydrolysis, buffered by PCr. Yet the quantitative interpretation stumbles on a common misconception that localizes the energy to only the intrinsic property of ATP. Eisenberg and Hill have aptly pointed out:

'The problem with treating ATPase systems as if they were energized by ATP is that the free energy of ATP hydrolysis is not localized in the ATP molecule. Nor does it arise simply from a change in molecular structure when ATP is hydrolyzed to ADP and P_i' (Eisenberg and Hill, 1985). No free energy change actually occurs when ATP hydrolyzes to ADP and P_i at the active site of myosin (Bagshaw and Trentham, 1973). Instead, the energy of ATP hydrolysis originates from the free energy difference of ATP, ADP and P_i in solution.

Nichols and Feruguson (1992) have re-emphasized these rigorous views of thermodynamics. They lambast the 'myth of the "high energy phosphate"' bond in ATP as the reservoir of stored energy, capable of driving reactions forward in energetically unfavorable conditions. Only the displacement of the mass action ratio from the equilibrium defines the capacity of ATP to do work, such that the shift in the mass–action ratio produces the driving force towards equilibrium or the free energy to do work ($\Delta G = \Delta G^{\circ} + RT \ln K$), where K is the mass action ratio. In the cell, the poorly defined K and the variability of K across fiber types pose a nettlesome obstacle in determining quantitatively the available ATP free energy.

In essence Hill set the framework to analyze the transduction of chemical energy into mechanical energy and provided key insight into the molecular mechanism of cross-bridge movement, but his analysis neither specified the available free energy nor clarified the rules directing cellular energy flow during muscle contraction.

8.2.2. Physiological Models

In early studies of muscle bioenergetics, much of the effort focused on devising techniques to detect and correlate rapid changes in metabolite, heat and work. Muscle physiologists/biochemists responded to Hill's exhortation to prove that ATP provided the energy source for cross-bridge movement during a twitch. Given that experimental perspective, the frog sartorius muscle at 0 °C served as an appropriate model.

The physiology of isolated frog sartorius muscle at 0 °C, however, differs significantly from normothermic mammalian striated muscle. Indeed Hill himself showed that a key premise, the Fenn effect or the heat of shortening, which initially launched his investigation into the viscoelastic model, didn't even materialize in gastrocnemius muscle (Hill, 1913). Other researchers have corroborated these anomalous findings and understandably have objected to exalting the Fenn effect as a universal paradigm (Woledge, 1971).

Moreover, the accepted energy per twitch value arises from snap-frozen isolated muscle at 0 °C, poisoned with iodoacetate, anoxic, containing inhibitors of creatine kinase and an uncoupler of respiration (Wilkie, 1968). Certainly these metabolic perturbations inhibited the major ATP generation pathways to simplify and narrow the evidence gathering to ascribe ATP as the fuel source for muscle contraction and PCr as the energy buffer, but these unphysiological perturbations also raise questions about the actual energy utilization and flow in normothermic striated muscle *in vivo* (Cain *et al.*, 1962a; Infante *et al.*, 1965; Mommaerts, 1954).

Even for frog sartorius muscle, the resting metabolic energy consumption at 20 °C vs 0 °C differs by a factor 8. Extrapolating to 40 °C would yield a factor of 16. In addition, the Ca^{2+} reuptake in the sarcoplasmic reticulum has a temperature-dependent energy increase of 2.5 for every increase of 10 °C (Q_{10}). So, from 0 to 40 °C, the energy change is approximately 10. Although, Wilkie's experimental results showing a linear relationship between heat + work and ΔPCr, the extent of ATP utilization per twitch in normothermic striated muscle actually remains unsettled (Wilkie, 1968).

8.2.3. Energy Cost per Twitch

The ambiguity in the quantitative determination of ΔPCr/twitch muscle arises from the limitations and the underlying assumptions of the two predominant experimental approaches, freeze-clamping and time-averaging.

With the freezing-clamping method, the time resolution depends critically on sample thickness and subsequent metabolic assay of the total pool, and has a time resolution of 100 ms (Gilbert *et al.*, 1971; Infante and Davies, 1962; Infante *et al.*, 1965; Wilkie, 1968). Yet, the time to maximum force development in a muscle contraction cycle completes in less than 20 ms. The freeze-clamp method does not have the time resolution to accurately monitor the energetic fluctuation of a twitch.

Understandably, the freeze-clamp data have led to debates surrounding the time course of ATP utilization. Frog sartorius muscle at 0 °C releases approximately 2 μmol/g muscle of P_i, which appears within 5 s of isometric tetanus. However, during the first 600 ms of tetanus, the average PCr splitting amounts to about 1 μmol/g tissue/s and breaks down nonlinearly, reacting faster in the initial phase than the later phase. In fact, muscle produces initially heat much faster than observed rate of PCr breakdown (Kushmerick and Davies, 1969; Kushmerick and Paul, 1975).

Photolysis experiments have also measured the ATP utilization rate in isolated muscle fiber. He *et al.* (1997) created a high-affinity phosphate binding protein that displayed a 5-fold increase in fluorescence upon phosphate binding. Intercalated into the isolated muscle fibers, the molecule monitored rapid P_i release during the onset of muscle contraction and reflected, according to the investigators' interpretation, a stoichiometric index of ATP hydrolysis. The cell also contained caged ATP molecules, which upon laser photolysis immediately liberated 1.5 mM of ATP. Consequently, a temporal response of the myosin ATPase to the sudden release of ATP yielded the kinetics profile of ATP hydrolysis. The initial rate of P_i released, or ATP hydrolysis, occurred 10 times faster than the corresponding steady-state rate (He *et al.*, 1998, 1999, 2000; Potma *et al.*, 1994, 1995). In particular, soleus and psoas muscle in the presence of Ca^{2+} showed an initial rate of P_i release of 0.57 and 4.7 mM/s. These disparate observations and experimental limitations of the freeze-clamp method argue against any consensus view of the interaction of ΔPCr/twitch, heat and work.

Alternatively, investigators have extrapolated the cost per twitch in rat muscle by following the PCr decline in 30 s data blocks during a series of stimulated contractions over 8 min. The signal from each block represents an averaged value at 15 s, and a curve fit to the monoexponential function $\{\mathrm{PCr}(t) = \mathrm{PCr(ss)} + [\mathrm{PCr}(0) - \mathrm{PCr(ss)}]\exp^{-t/\tau}\}$ yields the time constant of the PCr kinetics. If, as assumed, neither glycolysis nor respiration contributes significantly to the formation of ATP during each contraction interval, then the first derivative of the exponential function, evaluated at $t = 0$ will give the ΔPCr/twitch or ΔE/twitch (Meyer, 1988; Taylor *et al.*, 1986). In essence, the PCr kinetics analysis does not need to account for any creatine kinase back-reaction from ATP to PCr. The extrapolated phosphagen cost for rat muscle is about 0.25 μmol/g/twitch. For human gastrocnemius muscle, ΔPCr/twitch is 0.15 μmol/g/twitch (Blei *et al.*, 1993).

The analysis depends upon an averaged ΔPCr/twitch and assumes that the experimental observation over an averaged vs a transient period does not introduce any significant errors. However the time to peak force development is about 20 ms or 2 % of the total observation time under 1 Hz stimulation. Sampling the PCr change and then dividing by total number of twitches might mask a large, but transient fluctuation, which would undermine the validity of the energy balance analysis.

8.3. CURRENT PERSPECTIVES

8.3.1. NMR Measurements of Metabolic Transients

Many ^{31}P NMR studies have investigated the steady-state high-energy phosphate signals in normothermic striated muscle *in vivo* to map energy regulation *in vivo* (Kushmerick and Meyer, 1985; Kushmerick *et al.*, 1992a; Shoubridge and Radda, 1987). Measuring directly the ΔPCr/twitch, however, has faced formidable experimental hurdles (Foley and Meyer, 1993; Shoubridge *et al.*, 1984). With NMR, the low signal sensitivity precludes simple transient observations, since a single ^{31}P free induction decay has insufficient signal-over-noise ratio to yield a detectable NMR peak during a muscle contraction cycle. Certainly signal averaging will produce a measurable ^{31}P signal but will yield only an averaged time response that requires key assumptions to extrapolate to the energy cost per twitch.

A recent gated ^{31}P NMR technique with sufficient time resolution to monitor the PCr hydrolysis during a contraction cycle in stimulated rat gastrocnemius muscle shows that PCr level falls rapidly with a $t_{1/2}$ of 8 ms and exhibits a kinetics curve that mirrors the force development profile (Chung *et al.*, 1998). In contrast to previous measurements of about 0.15–0.3 mM ΔPCr/twitch, the gated NMR experiments show a ΔPCr/twitch drop of 3 mM during the contraction phase, and a PCr restoration to baseline level in the recovery phase. P_i changes stoichiometrically, while ATP level remains constant. The data indicate that PCr hydrolysis proceeds much more rapidly as well as extensively than previously observed or presumed.

Chung *et al.*'s (1998) results demonstrate that the gated NMR technique can detect with ±1 ms time resolution the pronounced PCr fluctuation during a contraction cycle (Table 8.1; Figures 8.4 and 8.5; Chung *et al.*, 1994). Figure 8.4 clearly shows that high-energy phosphate signal intensities change sharply: PCr falls, P_i rises stoichiometrically, and ATP remains constant. Such a distinct pattern of signal response argues strongly against a significant contribution from motional artifact as the muscle contracts. Moreover, any

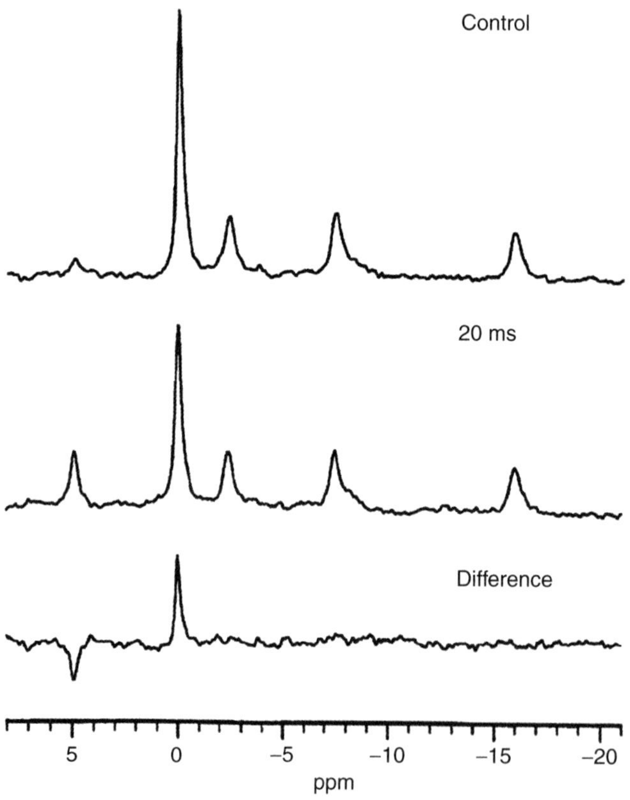

Figure 8.4. ^{31}P NMR spectra during a contraction cycle. The experiments show that a transient fluctuation in the PCr level during a contraction cycle appears in the ^{31}P spectra from rat gastrocnemius muscle: at rest the control ^{31}P spectrum from muscle displays the high-energy phosphate signals corresponding to PCr, P_i and ATP (top). Signals acquired at 20 ms distal to the sciatic nerve stimulation reveal a PCr signal intensity drop, a P_i signal increase, and a constant ATP level (middle). The difference spectrum shows clearly the intensity and direction of the ^{31}P signal response at 20 ms after stimulation (bottom) (Chung *et al.*, 1998).

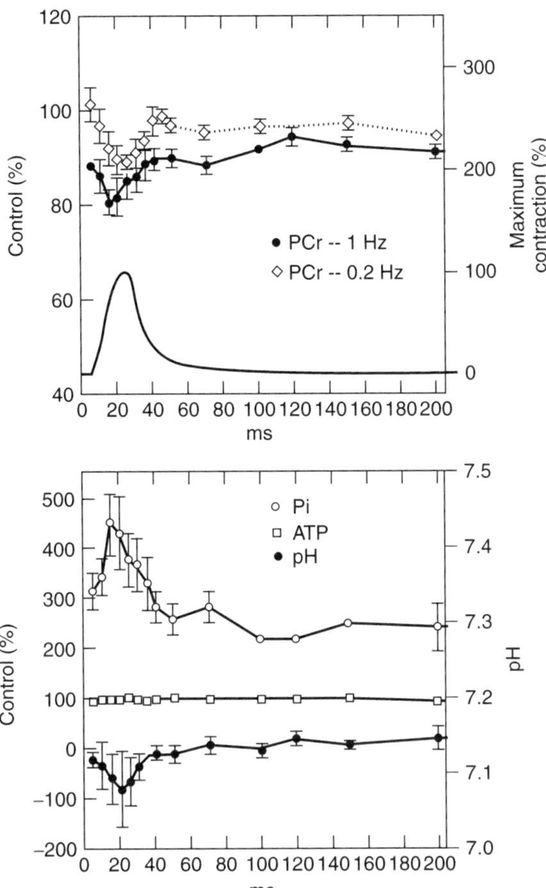

Figure 8.5. Metabolic transients during a twitch. During a twitch the PCr level declines about 10 %. The dynamic PCr profile is a slightly time-shifted mirror image of the force contraction response, which exhibits a maximum at 25 ms. The P_i rises about 1.8-fold and is inversely related to the PCr response. Based on the reported resting state values of PCr and P_i of 27.1 and 2.8 μmol/g wet tissue in rat gastrocnemius muscle, the calculated changes in PCr and P_i, after saturation factor correction, are 3.1 and 3.2 μmol/g wet tissue, respectively (Chung *et al.*, 1998). The observed PCr to P_i reaction appears to be stoichiometric. Concomitantly, pH appears to shift by 0.06 pH units, from 7.14 ± 0.02 to 7.08 ± 0.11. However the limited accuracy of the data measurement does not indicate that the change is statistically significant and precludes at this time any definitive interpretation. Dynamic physiological and metabolic response during a twitch: (a) Under 1 Hz stimulation, a single contraction profile shows a rapid rise in force development, which reaches a maximum at 25 ms. The force profile is a representative one, not the average of five experiments. Averaged force parameters are described in Table 8.1. At the same time the PCr level drops to a minimum at 16 ms (–). ATP level remains constant. A similar PCr profile is observed under 0.2 Hz stimulation (. . .). (b) Corresponding P_i and pH profiles indicate a stoichiometric increase in P_i during the contraction cycle. No statistically significant change in pH appears.

detection volume shift between fast and slow twitch fibers would produce a contrasting set of ^{31}P spectra, since in the transition from fast to slow fibers, PCr will decrease from 35 to 17 mM, P_i will increase from 3 to 10 mM, but ATP will decrease from 9 to 5 mM (Kushmerick *et al.*, 1992b; Madapillimattam *et al.*, 1994; Phillips *et al.*, 1993).

Table 8.1. Metabolic changes and contractile parameters during a muscle twitch. The control values originate from fully relaxed ^{31}P spectra. Steady-state values are mean percentages of control ± standard deviation (SD) of the data recorded at 100, 120, 150, 200, 500 and 900 ms time points. Maximal change is relative to the steady-state value. The corresponding relative maximum change occurs as follows: PCr, 16 ms; P_i, 16 ms; ADP, 16 ms; pH, 21 ms; ATP, 200 ms. Control pH = 7.11. Significant change is noted as follows: *paired t-test, $p<0.05$. All other reported changes are insignificant.

	Steady state		Transient (maximum change point)	
		n		*n*
Metabolic parameters				
PCr	92.1 % ± 3.0	17	80.8 % ± 5.8*	5
ATP	98.1 % ± 5.9	16	95.2 % ± 3.2	5
P_i	248.9 % ± 54.5	17	447.3 % ± 138.3*	5
ADP	41.4 % ± 3.8	16	51.0 % ± 12.1	5
pH	7.14 ± 0.02	17	7.08 ± 0.11	5
Contractile parameters				
Time to maximum peak tension	24.5 ± 3.8 ms			8
Half relaxation time	10.5 ± 2.8 ms			8
Peak force	1.10 ± 0.08 g/g body weight ± SD			8

8.3.2. ΔPCr/Twitch

During a muscle twitch, the $t_{1/2}$ for the PCr kinetics is 8 and 14 ms, for the respective stimulation and recovery phase. The values correspond to first-order rate constants of 87 or 50 s^{-1} and PCr formation and degradation rates of 182 and 106 μmol/g tissue/s, respectively. From the steady state level, PCr falls by 11.3 %, corresponding to a decrease of 3 μmol ATP/g tissue per twitch (Chung *et al.*, 1994). The change in PCr is equivalent to a 40 % turnover of the total ATP content (London, 1991; Meyer *et al.*, 1985).

Using the non-gated steady-state data from the same experiment and extrapolating to dPCr/dt at $t = 0$ yields only a 0.3 % turnover or 0.08 μmol/g tissue, within the range of previously reported steady-state values (Chung *et al.*, 1994; Curtin and Woledge, 1978; Foley and Meyer, 1993; Homsher and Kean, 1982). The energy cost per twitch derived from steady-state analysis clearly underestimates the ΔPCr/twitch.

8.3.3. Creatine Kinase

The metabolic energy cost per twitch appears quite large and raises questions about the prevailing paradigm about PCr mobilization and ATP utilization during a muscle contraction. For the PCr pool to buffer the ATP utilization at 182 μmol/g tissue/s, the creatine kinase reaction must have the capacity to shift dramatically from its resting state flux of 7.4 μmol/g tissue/s from PCr to ATP. Yet the enhanced forward flux from PCr to ATP is well below the estimated V_{max} of 224 μmol/g tissue/s (Meyer *et al.*, 1986). Given the resting PCr forward flux rate of 7.4 μmol/g tissue/s in fast twitch muscle, the creatine kinase can accommodate the increased forward flux as the ADP concentration rises from its resting value of 0.016 mM by a factor of 26 during a contraction.

8.4. IMPLICATIONS

8.4.1. Energetics of Muscle Contraction

The current paradigm envisions a low energy cost per twitch and an insignificant ATP restoration during the contraction cycle. Instead of a ΔPCr/twitch of 0.15–0.3 mM of ATP, a 3 mM ATP consumption per contraction challenges the paradigm and requires another fuel source to buffer the millisecond energy need during the contraction-recovery cycle.

8.4.2. Oxygen Consumption

The muscle literature embraces a ΔPCr/twitch about 0.15–0.25 mM ATP in normothermic striated muscle, in agreement with ΔPCr/twitch of 0.36 μmol/g muscle (~0.3 mM) in metabolically inhibited frog sartorius muscle at 0 °C. Each contraction does not demand significant energy and therefore does not need any ATP resynthesis (Cain *et al.*, 1962a; Foley and Meyer, 1993; Infante *et al.*, 1965). Researchers have used the PCr kinetics during the recovery period after muscle contraction has ceased to indicate a slow rate for oxidative phosphorylation (Kushmerick *et al.*, 1992a). In such analysis, the time constant for PCr kinetics during contraction is 1.32 min, whereas the PCr time constant for the recovery is 1.57 min (Meyer, 1989). Converting the dPCr/d*t* to VO_2 assumes a uniformly distributed creatine kinase at instantaneous equilibrium with the reactants, a non-limiting O_2 and substrate supply, a negligible anaerobic ATP contribution, a constant P/O and basal VO_2, and a predominant uni-directional creatine kinase reaction that consumes all the resynthesized ATP (Foley and Meyer, 1993; Meyer, 1989). In particular, if the last two assumptions fail to hold, then the analysis would significantly underestimate the extrapolated VO_2.

Indeed, the VO_2 values obtained from arterial–venous O_2 difference measurements and dividing the steady-state VO_2 by the number of contractions encompass a broad range of values and depend highly upon the experimental model and conditions. Isolated cat biceps and soleus consume approximately 1.9–3.3 μM O_2 per contraction, respectively (Kushmerick *et al.*, 1992a). Perfused rat hindlimb utilizes 37 μM O_2 per contraction or 260–350 μM O_2 per tetanic contraction (Hood *et al.*, 1986; Robinson *et al.*, 1994; Ruderman *et al.*, 1980). The broad range of VO_2 values arises from variation in muscle aerobic capacity, mitochondrial content, temperature, experimental model and oxygen consumption measurement techniques.

8.4.3. Intracellular O_2 and VO_2

A key perspective on the bioenergetics of muscle contraction arises from the ^{1}H NMR studies that have mapped the intracellular PO_2 with the myoglobin (Mb) signals *in vivo* (Jue and Anderson, 1990; Kreutzer *et al.*, 1992; Mole *et al.*, 1999; Tran *et al.*, 1999). These NMR studies have followed the signal intensity of the proximal histidyl $N_\delta H$ signal of deoxy Mb and the γCH_3 Val E11 signal of oxyMb to yield an index of the intracellular PO_2 (Kreutzer *et al.*, 1992). In resting muscle, O_2 fully saturates MbO_2, since experiments do not detect any deoxy Mb signal in the spectra of skeletal muscle or rat myocardium (Chung *et al.*, 1998; Kreutzer *et al.*, 1992, 2001; Mole *et al.*, 1999; Zhang *et al.*, 1999). Because the sensitivity of NMR can detect a 10% deoxy Mb signal, the resting pO_2 must be greater than 20 mmHg but cannot exceed 40 mmHg, the venous pO_2 value. The observation implies that the myocyte PO_2 rests well above 2.93 mmHg (Mb p50 at 39 °C) and must also saturate cytochrome oxidase (Chance, 1989). If O_2 is the limiting molecule in regulating VO_2, then raising only the cellular PO_2 cannot increase cytochrome oxidase activity. Lowering the PO_2 should certainly decrease the VO_2. Yet, human gastrocnemius muscle studies clearly show a decrease in PO_2 as VO_2 increases with exercise, in agreement with the cryosection analysis of stimulated canine gracilis muscle (Gayeski and Honig, 1988; Mole *et al.*, 1999). The observation has provoked a continuing discussion on the actual contribution of O_2 to regulating respiration in contracting muscle.

The Mb desaturation kinetics provides a means to measure the VO_2 under the assumption that Mb has a primary function to deliver O_2 to the mitochondria and provides the predominant pool of O_2 at the initiation of contraction. The dynamic change in the ^1H NMR proximal histidyl NH signal of Mb reflects the rate of cellular demand for O_2. Indeed, at the beginning of muscle contraction, Mb desaturates rapidly with an exponential time constant of 30 s (Chung *et al.*, 2004). As contraction frequency increases at 45, 55 and 70 rpm with a constant load, the steady-state level of Mb desaturation increases from 0 to 30, 36 and 48 %, reflecting a progressive drop in intracellular PO_2 but an enhanced O_2 gradient from the vasculature into the cell (Mole *et al.*, 1999). The desaturation kinetics of MbO_2 implies that the cell draws primarily from the Mb oxygen at the start of muscle contraction (Wittenberg and Wittenberg, 1989). The rate of Mb desaturation reflects then the intracellular VO_2.

Given the cellular Mb concentration of approximately 0.4 mM in human gastrocnemius muscle and a resting VO_2 of 2.02 μM/s, the dMb/dt yields the intracellular VO_2 ranging from 7.5 to 8.9 μM O_2/s at the onset of contraction (Blei *et al.*, 1993). The observed VO_2 corresponds to the NIRS determined VO_2 of 3.0–31.2 μM O_2/s in forearm muscle during isometric handgrip exercise. The broad range of NIRS values may arise from a composite Mb and Hb contribution to the signal (Sako *et al.*, 2000; Tran *et al.*, 1999).

Oxidative phosphorylation also requires an available ADP pool. For glycolytic muscle, 0.5 Hz stimulation will raise the calculated ADP level from 1.3 to 65 μmol/g tissue, a factor of 50, whereas for oxidative fiber the ADP level will rise from 4.1 to 30 μmol/g tissue, a factor of 7.3 (Kushmerick *et al.*, 1992a). The dramatic shift in the ADP levels still remains within the ADP translocase K_m of 6–66 μmol/g tissue (Jacobus *et al.*, 1982; Kushmerick *et al.*, 1992a). During muscle contraction VO_2 can then increase by a factor of 20, while ADP can increase by a factor of 50.

8.4.4. VO_2-dependent ATP Production during Muscle Contraction

The Mb desaturation kinetics indicates an intracellular consumption of 7.5–8.9 μM O_2/s at the onset of contraction. Using the canonical P/O ratio of 3/1 yields the corresponding oxidative ATP production rate of 45–54 μM/s or approximately 50 μM per contraction. Given the 0.36 mM (~0.30 mM) ATP/g per contraction observed in freeze-clamp experiments with frog muscle, oxidative ATP production can supply 17 % of the ATP required at the initiation of contraction (Infante *et al.*, 1965). Steady-state NMR measurement of the PCr signal from human gastrocnemius muscle stimulated at 1 Hz, however, has yielded an energy cost of only 0.15 mM ATP per contraction (Blei *et al.*, 1993). Oxidative ATP production can supply up to 33 % of the total energy and represents a significant fraction of the currently accepted view of the energy cost per muscle contraction.

With the consumption of 3 mM ATP per contraction, however, the intracellular VO_2, as determined by the Mb desaturation kinetics, implies that oxidative phosphorylation can only supply 2 % of the energy per twitch (Chung *et al.*, 2004). PCr would deplete rapidly, unless an anaerobic energy process restored the PCr level during contraction. Although PCr is the immediate energy source for ATP formation, glycolysis and glycogenolysis can provide ATP during muscle contraction (Spriet, 1992). A glycogen shunt theory has now proposed that the ~70 mM glucosyl unit in the glycogen pool supplies that energy source for the millisecond contraction need (Shulman and Rothman, 2001b). In effect, it proposes a quantitative link between glycogenolysis and the temporal energy demand during a muscle contraction cycle.

Glycogen, however, will support a limited number of contractions at a ΔPCr/twitch corresponding to 3 mM of ATP, unless an additional energy replenishment shifts the creatine kinase flux to restore PCr. Otherwise, glycogen will completely deplete after several dozen contractions, which experiments have not observed. Such energy restoration must involve oxidative phosphorylation at the steady state.

The initial VO_2 does not appear sufficient to meet the initial contraction energy need and lends support for glycogen as the anaerobic fuel source. However, the intracellular VO_2 analysis assumes that Mb supplies all the O_2 and that free O_2 does not contribute until Mb has reached its deoxygenated steady state when the vasculature has adapted the blood flow to deliver more O_2. The intracellular VO_2 value, derived from the Mb desaturation kinetics, represents only a lower limit. Indeed, VO_2 continues to increase even after Mb desaturation has reached a steady-state level. The increasing VO_2 must depend upon the additional O_2 from the enhanced blood flow. Indeed, the maximum VO_2 can exceed the basal VO_2 rate by 30 times. Even though the underestimated VO_2/contraction value derived from Mb desaturation kinetics does not appear sufficient to provide ATP to replenish PCr at the onset of contraction, the increasing VO_2 at the steady state can provide significant ATP as muscle contraction proceeds. Such an interplay of VO_2 and energy restoration requires a much more significant role of oxidative phosphorylation during muscle contraction than previously envisioned.

8.5. INTEGRATIVE VIEW

The transduction of mechanical, thermal and chemical energy during muscle contraction stands as a central paradigm in the field of bioenergetics. Over many decades, physiologists have composed a view of the molecular interaction and energy regulation, based largely on steady-state measurements. These steady-state measurements lead to an extrapolated energy cost/utilization during a muscle contraction cycle and assert the insignificant role for any metabolic transients, based on early freeze-clamp experiments. Yet the freeze-clamp techniques have insufficient time resolution to capture a single metabolic transient. Moreover, the unphysiological experimental models and the underlying assumptions in the analysis appear inadequate to explain the regulation of energy flow during a contraction cycle in normothermic, blood-perfused striated muscle. With the advent of *in vivo* NMR, new techniques emerge to interrogate directly reaction kinetics muscle. Indeed, Chung *et al.* (1998) developed a gated NMR technique to observe the transient fluctuation in PCr during a contraction cycle and noted that PCr decreased much more than previously reported or expected. Their reported value of 3 mM ATP/twitch implies that a dynamic ATP synthesis and degradation must occur during each muscle transient.

To discriminate between the different ATP synthesis pathways requires a definitive determination of the critical intracellular VO_2. Current techniques extrapolate the intracellular VO_2 value from arterial venous O_2 difference and whole body VO_2 measurements. Recently, Chung *et al.* (2004) have utilized the Mb desaturation kinetics in exercising muscle to determine the intracellular VO_2 at the onset of contraction. Based on the intracellular VO_2, oxidative phosphorylation can only supply 2 % of the required ATP/twitch.

The limited energy supply from oxidative phosphorylation at the onset of contraction and the high energy demand for each contraction requires then a rapid anaerobic ATP synthesis pathway to maintain muscle contraction. Shulman and Rothman (2001b) have proposed that glycogen supplies the fuel during the rapid, millisecond period of a muscle contraction. Between contractions glycogen must regenerate ATP in milliseconds to replenish PCr. This glycogen shunt mechanism takes advantage of the rapid activation of glycogen phosphorylase and allows lactate to serve as a time buffer between the fast and slow energy needs.

Oxidative phosphorylation must still play a pivotal role, since the glycogen depot can support only a limited number of contractions. Even though the initial intracellular VO_2 supplies only 2 % of the energy need, VO_2 is not static. Indeed, VO_2 continues to rise with muscle contraction as the vasculature and blood flow adapt to the increased energy demand. At the steady state then, oxidative phosphorylation can provide much more ATP to restore the energy pool.

In 1950 Hill challenged biochemists to prove muscle consumes ATP during a twitch. That call launched a scientific inquiry on muscle bioenergetics, which then defined the research field. In the intervening years, new scientific findings and techniques have emerged and have compelled investigators to re-examine some of these defining principles in light of metabolic transients. In that enduring scientific spirit of discovery, Shulman has now issued a similar exhortation with his glycogen shunt theory: show that glycogenolysis, balanced by oxidative phosphorylation, governs the transient, dynamic energy flow to support ATP utilization during muscle contraction.

Acknowledgments

I would like to acknowledge the funding support from NIH GM58688, NSF MBCE 0077595 and Phillip Morris 005510, and the delightful scientific discussions with Drs Youngran Chung, Ulrike Kreutzer Paul Mole, Thomas Gayeski, Richard Connett and Marty Kushmerick, who have shaped my perspectives about the elegant rules governing muscle metabolism and physiology. A special note of appreciation goes to Bob Shulman. Without Bob's engaging conversation, insightful comments, careful editing and patient but persistent goading, the chapter would never have materialized.

REFERENCES

Aidley DJ (1989) *The Physiology of Excitable Cells*. Cambridge: Cambridge University Press.

Alger JR and Shulman RG (1984) NMR studies of enzymatic rates *in vitro* and *in vivo* by magnetization transfer. *Q Rev Biophys* **17**: 83–124.

Bagshaw CR and Trentham (1973) The reversibility of adenosine triphosphate cleavage by myosin. *Biochem J* **133**: 323–328.

Blei ML, Conley KE and Kushmerick MJ (1993) Separate measures of ATP utilization and recovery in human skeletal muscle. *J Physiol* **465**: 203–222.

Cain DF, Infante AA and Davies RE (1962a) Adenosine triphosphate and phosphorylcreatine and energy supplies for single contractions of working muscle. *Nature* **196**: 214–217.

Cain DF, Infante AA and Davies RE (1962b) Chemistry of muscle contraction. *Nature* **196**: 214–217.

Chance B (1989) Metabolic heterogeneities in rapidly metabolizing tissue. *J Appl Cardiol* **4**: 207–221.

Chance B and LaNoue KF (1990) Metabolic control in exercising skeletal muscle. *Am J Physiol* **258**: R288–R290.

Chance B, Eleff S, Leigh JS, Sokolow DP and Sapega AA (1981) Mitochondrial regulation of phosphocreatine/inorganic phosphate ratios in exercising human muscle: a gated ^{31}P NMR study. *Proc Natl Acad Sci USA* **78**: 6714–6718.

Chung Y, Sharman R, Carlsen R, Unger S, Larson D and Jue T (1994) Metabolic fluctuation in skeletal muscle during a muscle contraction cycle. *Proc Soc Magn Reson Med* **1**: 360.

Chung Y, Sharman R, Carlsen R, Unger SW, Larson D and Jue T (1998) Metabolic fluctuation during a muscle contraction cycle. *Am J Physiol* **274**: C846–C852.

Chung Y, Mole P, Sailasuta N, Tran TK, Hurd R and Jue T (2004) Control of respiration and bioenergetics during muscle contraction. *Am J Physiol* (in press).

Connett RJ and Honig CR (1989) Regulation of VO2 in red muscle: do current biochemical hypotheses fit *in vivo* data? *Am J Physiol* **256**: R898–R906.

Connett RJ and Honig CR (1990) Metabolic control in exercising skeletal muscle. *Am J Physiol* **258**: R288–R290.

Cozzone PJ and Bendahan D. (1994) ^{31}P NMR spectroscopy of metabolic changes associated with muscle exercise: physiopathological applications. In: *NMR in Physiology and Biomedicine*, edited by Gillies RJ. San Diego, CA: Academic Press; 389–402.

Curtin NA and Woledge RC (1978) Energy changes and muscular contraction. *Physiol Rev* **58**: 690–716.

Dawson MJ, Gadian DG and Wilkie DR (1977) Contraction and recovery of living muscles studied by ^{31}P nuclear magnetic resonance. *J Physiol* **267**: 703–735.

Eggleton P and Eggleton GP (1927) The inorganic phosphate and a labile form of organic phosphate in the gastrocnemius muscle of the frog. *Biochem J* **21**: 190–195.

Eisenberg E and Hill TL (1985) Muscle contraction and free energy transduction in biological systems. *Science* **227**: 999–106.

Fenn WO (1923) A quantitative comparison between the energy liberated and the work performed by the isolated sartorius of the frog. *J Physiol (Lond)* **58**: 175–203.

Fenn WO (1924) The relation between the work performed and the energy liberated in muscular contraction. *J Physiol* **58**: 373–395.

Fiske CH and Subbarrow Y (1927) The nature of the 'inorganic phosphate' in voluntary muscle. *Science* **65**: 401–403.

Fletcher WM and Hopkins FG (1907) Lactic acid in amphibian muscle. *J Physiol* **35**: 247–309.

Foley JM and Meyer RA (1993) Energy cost of twitch and tetanic contractions of rat muscle estimated *in situ* by gated ^{31}P NMR. *NMR Biomed* **6**: 32–38.

Gage PW and Eisenberg RS (1969) Action potentials, after potentials, and excitation–contraction coupling in frog sartorius fibers without transverse tubules. *J Gen Physiol* **53**: 265–278.

Gayeski TEJ and Honig CR (1988) Intracellular Po_2 in long axis of individual fiber in working dog gracilis muscle. *Am J Physiol* **254**: H1179–H1185.

Gilbert C, Kretzschmar KM, Wilkie DR and Woledge RC (1971) Chemical change and energy output during muscular contraction. *J Physiol* **218**: 163–193.

He ZH, Chillingworth RK, Corrie JET, Trentham DR, Webb MR and Ferenczi MA (1997) ATPase kinetics on activation of rabbit and frog permeabilized isometric muscle fibres: a real time phosphate assay. *J Physiol* **501**: 125–148.

He ZH, Stienen GJM, Barends JPF and Ferenczi MA (1998) Rate of phosphate release after photoliberation of adenosine 5′-triphosphate in slow and fast skeletal muscle fibers. *Biophys J* **75**: 2389–2401.

He ZH, Chillingworth RK, Brune M, Corrie JET, Webb MR and Ferenczi MA (1999) The efficiency of contraction in rabit skeletal muscle fibres, determined from the rate of release of inorganic phosphate. *J Physiol* **517**: 839–854.

He ZH, Bottinelli R, Pellegrino MA, Ferenczi MA and Reggiani C (2000) ATP consumption and efficiency of human single muscle fibers with different myosin isoform composition. *Biophys J* **79**: 945–961.

Hill AV (1913) The absolute mechanical efficiency of the contraction of an isolated muscle. *J Physiol* **46**: 435–469.

Hill AV (1938) The heat of shortening and the dynamic constants of muscle. *Proc R Soc Lond B* **126**: 136–195.

Hill AV (1950) A challenge to biochemists. *Biochim Biophys Acta* **4**: 4–11.

Hill AV. (1965) *Trails and Trials in Physiology*. London: Arnold.

Homsher E and Kean CJC (1982) Unexplained enthalpy production in contracting skeletal muscles. *Fed Proc* **41**: 149–154.

Homsher E, Irving M and Wallner A (1981) High-energy phosphate metabolism and energy liberation associated with rapid shortening in frog skeletal muscle. *J Physiol* **321**: 423–436.

Hood DA, Gorski J and Terjung RL (1986) Oxygen cost of twitch and tetanic isometric contractions of rat skeletal muscle. *Am J Physiol* **250**: E449–E456.

Huxley HE and Hanson J (1954) Changes in the cross-striation of muscle during contraction and stretch and their structural interpretation. *Nature* **173**: 973–976.

Huxley AF and Niedergerke R (1954) Structural changes in muscle during contraction. Interference microscopy of living muscle fibres. *Nature* **173**: 971–973.

Huxley AF and Simmons RM (1971) Proposed mechanism of force generation in straited muscle. *Nature* **233**: 533–538.

Huxley AF and Simmons RM (1973) Mechanical transients and the origin of muscular forces. *Cold Spring Harbor Symp Quant Biol* **37**: 669–680.

Huxley AF and Taylor RE (1958) Local activation of striated muscle fibres. *J Physiol* **144**: 426–441.

Iles RA, Stevens AN, Griffiths JR and Morris PG (1985) Phosphorylation status of the liver by ^{31}P nmr spectroscopy and its implications for metabolic control. *Biochem J* **229**: 141–151.

Infante AA and Davies RE (1962) Adenosine triphosphate breakdown during a single isotonic twitch of frog sartorius muscle. *Biochem Biophys Res Commun* **9**(5): 410–415.

Infante AA, Klaupiks D and Davies RE (1965) Phosphorylcreatine consumption during single-working contractions of isolated muscle. *Biochim Biophys Acta* **94**: 504–515.

Jacobus WE, Moreadith RW and Vandegaer KM (1982) Mitochondrial respiratory control. *J Biol Chem* **257**: 2397–2402.

Jue T and Anderson S (1990) ^{1}H observation of tissue myoglobin: an indicator of intracellular oxygenation *in vivo*. *Magn Reson Med* **13**: 524–528.

Kretzschmar KM and Wilkie DR (1969) A new approach to freezing tissues rapidly. *J Physiol* **202**: 66P–67P.

Kreutzer U, Wang DS and Jue T (1992) Observing the ^{1}H NMR signal of the myoglobin Val-E11 in myocardium: an index of cellular oxygenation. *Proc Natl Acad Sci USA* **89**: 4731–4733.

Kreutzer U, Mekhamer Y, Chung Y and Jue T (2001) Oxygen supply and oxidative phosphorylation limitation in rat myocardium *in situ*. *Am J Physiol Heart Circul Physiol* **280**: H2030–H2037.

Kushmerick MJ. (1983) Energetics of muscle contraction. In: *Skeletal Muscle*, edited by Peachey LD. Bethesda, MD: American Physiological Society; 189–236.

Kushmerick MJ and Davies RE (1969) The chemical energetics of muscle contraction. II. The chemistry, efficiency and power of maximally working sartorius muscles. *Proc R Soc Lond B* **174**: 315–353.

Kushmerick MJ and Meyer RA (1985) Chemical changes in rat leg muscle by phosphorus nuclear magnetic resonance. *Am J Physiol* **248**: C542–C549.

Kushmerick MJ and Paul RJ (1975) Relationship between initial chemical reactions and oxidative recovery metabolism for single isometric contractions of frog sartorius at 0 °C. *J Physiol* **254**: 711–727.

Kushmerick MJ, Meyer RA and Brown TR (1992a) Regulation of oxygen consumption in fast- and slow-twitch muscle. *Am J Physiol* **263**: C598–C606.

Kushmerick MJ, Meyer RA and Brown TR (1992b) Regulation of oxygen consumption in fast- and slow-twitch muscle. *Am Physiol Soc* **263**: C598–C606.

Levin A and Wyman J (1927) The viscous elastic properties of muscle. *Proc R Soc Lond B* **101**: 218–243.

Lipmann F (1941) Metabolic generation and utilization of phosphate bond energy. *Adv Enzymol* **1**: 99–162.

Lohmann K (1934) Uber die enzymatische Aufspaltung der Kreatinphosphorsaure: zugleich enin Beitrag zum Chemismus der Muskelkontraktion. *Biochem Z* **271**: 264–277.

London RE (1991) Methods for measurement of intracellular magnesium: nmr and fluorescence. *A Rev Physiol* **53**: 241–258.

Lundsgaard E (1930a) Untersuchungen uber Muskelkontraktionen ohne Milchsaurebildung. *Biochem Z* **217**: 162–177.

Lundsgaard E (1930b) Weitere Untersuchungen uber Muskelkontraktionen ohne Milchsaurebildung. *Biochem Z* **227**: 51–83.

Madapillimattam AG, Cross A, Nishio ML and Jeejeehoy KN (1994) Stability of high-energy substrates in fast- and slow-twitch muscle: comparison of enzymatic assay of biopsy with *in vivo* ^{31}P nuclear magnetic resonance spectroscopy. *Anal Biochem* **217**: 103–109.

Matthews PM, Bland JL, Gadian DG and Radda GK (1982) A ^{31}P-NMR saturation transfer study of the regulation of creatine kinase in the rat heart. *Biochim Biophys Acta* **721**: 312–320.

McComas AJ. (1996) *Skeletal Muscle: Form and Function*. Champaign, IL: Human Kinetics.

McCully K, Vandenborne K, Posner JD and Chance B. (1994) Magnetic resonance spectroscopy of muscle bioenergetics. In: *NMR in Physiology and Biomedicine*, edited by Gillies RJ. San Diego, CA: Academic Press; 405–411.

Meyer RA (1988) A linear model of muscle respiration explains monoexponential phosphocreatine changes. *Am J Physiol* **254**: C548–C553.

Meyer RA (1989) Linear dependence of muscle phosphocreatine kinetics on total creatine content. *Am J Physiol Cell Physiol* **257**: C1149–C1157.

Meyer RA and Foley JM. (1996) Cellular processes integrating the metabolic response to exercise. In: *Exercise: Regulation and Intergration of Multiple Systems*, edited by Rowell LB and Shepherd JT. New York: Oxford University Press; 841–869.

Meyer RA, Kushmerick MJ and Brown TR (1982) Application of ^{31}P-NMR spectroscopy to the study of striated muscle metabolism. *Am J Physiol* **242**: C1–C11.

Meyer RA, Sweeney HL and Kushmerick MJ (1984) A simple analysis of the 'phosphocreatine shuttle'. *Am J Physiol* **246**: C365–C377.

Meyer RA, Brown TR and Kushmerick MJ (1985) Phosphorus nuclear magnetic resonance of fast- and slow-twitch muscle. *Am J Physiol* **248**: C279–C287.

Meyer RA, Brown TR, Krilowicz BL and Kushmerick MJ (1986) Phosphagen and intracellular pH changes during contraction of creatine-depleted rat muscle. *Am Physiol Soc* **250**: C264–C274.

Meyerhof O and Lohmann K (1932) Uber energetische Wechselbeziehungen zwischen dem Umstatz der Phosphosaure ester im Muskelextrakt. *Biochem Z* **253**: 431–461.

Mole PA, Chung Y, Tran TK, Sailasuta N, Hurd R and Jue T (1999) Myoglobin desaturation with exercise intensity in human gastrocnemius muscle. *Am J Physiol* **277**: R173–R180.

Mommaerts WFHM (1954) Is adenosine triphosphate broken down during a single muscle twitch? *Nature* **174**: 1083–1084.

Needham DM. (1971) *Machina Carnis*. Cambridge: Cambridge University Press.

Nichols DG and Ferguson SJ. (1992) Quantitative bioenergetics: the measurement of driving forces. In: *Bioenergetics 2*, San Diego, CA: Academic Press; 39–63.

Nunnally RL and Hollis DP (1979) Adenosine triphosphate compartmentation in living hearts: a phosphorus nuclear magnetic resonance saturation transfer study. *Biochemistry* **18**: 3642–3646.

Phillips SK, Takei M and Yamada K (1993) The time course of phosphate metabolites and intracellular pH compared to recovery in rat soleus muscle. *J Physiol* **460**: 693–704.

Potma EJ, van Graas IA and Steinen GJM (1994) Effects of pH on myofibrillar ATPase activity in fast and slow skeletal muscle fibers of the rabbit. *Biophys J* **67**: 2404–2410.

Potma EJ, van Graas IA and Steinen GJM (1995) Influence of inorganic phosphate and pH on ATP utilization in fast and slow skeletal muscle fibers. *Biophys J* **69**: 2580–2589.

Radda GK (1986) The use of NMR spectroscopy for the understanding of disease. *Science* **233**: 640–645.

Rees D, Smith MB, Harley J and Radda GK (1989) *In vivo* functioning of creatine phosphokinase in human forearm muscle studied by ^{31}P NMR saturation transfer. *Magn Reson Med* **9**: 39–52.

Robinson DM, Oglivie R, Tullson PC and Terjung RL (1994) Increased peak oxygen consumption of trained muscle requires increased electron flux capacity. *J Appl Physiol* **77**: 1941–1952.

Ruderman NB, Kemmer FW, Goodman MN and Berger M (1980) Oxygen consumption in perfused skeletal muscle. *Biochem J* **190**: 57–64.

Sako T, Hamaoka T, Higuchi H, Kurodawa Y and Katsumura T (2000) Validity of NIR spectroscopy for quantitatively measuring muscle oxidative metabolic rate in exercise. *J Appl Physiol* **90**: 338–344.

Shoubridge EA and Radda GK (1987) A gated ^{31}P NMR study of tetanic contraction in rat muscle depleted of phosphocreatine. *Am J Physiol* **252**: 532–542.

Shoubridge EA, Bland JL and Radda GK (1984) Regulation of creatine kinase during steady state isometric twitch contraction in rat skeletal muscle. *Biochim Biophys Acta* **805**: 72–78.

Shulman RG and Rothman DL (2001a) ^{13}C NMR of intermediary metabolism: implications for systemic physiology. *A Rev Physiol* **63**: 15–48.

Shulman RG and Rothman DL (2001b) The 'glycogen shunt' in exercising muscle: a role for glycogen in muscle energetics and fatigue. *Proc Natl Acad Sci USA* **98**: 457–461.

Spriet LL (1992) Anaerobic metabolism in human skeletal muscle during short-term, intense activity. *Can J Physiol Pharmac* **70**: 157–165.

Taylor DJ, Styles P, Matthews PM, Arnold DA, Gadian DG and Radda GK (1986) Energetics of human muscle: Exercise-induced ATP depletion. *Magn Reson Med* **3**: 44–54.

Tran TK, Sailasuta N, Kreutzer U, Hurd R, Chung Y, Mole P, Kuno S and Jue T (1999) Comparative analysis of NMR and NIRS measurements of intracellular PO_2 in human skeletal muscle. *Am J Physiol* **276**: R1682–R1690.

Ugurbil K (1985) Magnetization transfer measurements of individual rate constants in the presence of multiple reactions. *J Magn Reson* **64**: 207–219.

Wahler BE and Wollenberger A (1958) Determination of orthophosphate in the presence of phosphate compounds with an affinity for acids and molybdate. *Biochem Z* **329**: 508–520.

Wilkie DR (1968) Heat work and phosphorylcreatine break-down in muscle. *J Physiol* **195**: 157–183.

Wittenberg BA and Wittenberg JB (1989) Transport of oxygen in muscle. *A Rev Physiol* **51**: 857–878.

Woledge RC. (1971) Heat production and chemical change in muscle. In: *Progress in Biophysics and Molecular Biology*, edited by Butler JAV and Noble D. New York: Pergamon Press; 37–72.

Woledge RC (1973) *In vitro* calorimetric studies relating to the interpretation of muscle heat experiments. *Cold Spring Harbor Symp Quant Biol* **37**: 629–634.

Woledge RC, Curtin NA and Homsher E. (1985) *Energetic Aspects of Muscle Contraction*. New York: Academic Press.

Zhang J, Murakami Y, Zhang Y, Cho Y, Ye Y, Gong G, Bache R, Ugurbil K and From AHL (1999) Oxygen delivery does not limit cardiac performance during high work states. *Am J Physiol* **276**: H50–H57.

9

Lactate, Glycogen and Fatigue

Robert G. Shulman and Douglas L. Rothman

Departments of Molecular Biophysics and Biochemistry and Diagnostic Radiology, Yale University Medical School, MR Center, New Haven, CT 06520-8043, USA

The modern age of lactate studies began approximately 20 years ago when George Brooks questioned the accepted findings of previous generations and proposed the lactate shuttle (1). Early studies can be traced back to Fletcher and Hopkins who, in 1907, demonstrated that lactate accumulated when frog muscles contracted up to the point of fatigue but disappeared during recovery in the presence of oxygen. Several generations of great physiologists and biochemists built upon these results to create a canonical view of lactate that dominates textbooks to this day. In this view, lactate is created by exercising muscle due to the deficit of oxygen. The supposition is that the additional energy needed for muscle work exceeds the supply

Metabolomics by In Vivo NMR. Edited by R. G. Shulman and D. L. Rothman
 ISBN: 0-470-84719-0

of oxygen so that lactate, the product of anaerobic glycolysis, is formed and then cannot be further oxidized. Brooks extended this picture by demonstrating that lactate could be shuttled readily, via monocarboxylic acid transporters (MCT), to cellular regions where it could be further metabolized, thereby conserving the energy available from its reducing equivalents [for a review see (2)]. More recently, lactate shuttling from astroglia to neurons within the same region of brain has been proposed to form a crucial normal link in intracellular signaling and the energetics of glutamate neurotransmitter cycling (3, 4).

The finding that lactate could be used to redistribute energy led to questions about its origin, long assumed merely to be a mismatch between oxygen supply and demand. Numerous experiments by Brooks (5, 6) and others have shown that lactate, in fact, is generated during work by skeletal muscle in the *presence* of plentiful levels of oxygen (7, 8). Lactate production occurs in fully oxygenated contracting muscles in which the mitochondrial electron transport chain is not restricted by a lack of oxygen. In fact it is generated even when mitochondria are fully oxidized (7, 9). NMR studies of the oxygenation of muscle myoglobin in exercising humans showed that in working muscle oxygen levels decrease with load but even at maximum oxygen consumption are well above the mitochondrial needs (10). Further, glycogen depletion occurs under conditions of moderate, sustainable exercise, during which the amount of adenosine triphosphate (ATP) produced from glycolysis is small. However, exhaustion of glycogen and glucose under these conditions still leads to cessation of the ability to perform the exercise.

Studies of brain by functional imaging have shown that local plasma oxygen concentration *increases* during periods of increased neural activity (11). Yet despite the availability of abundant oxygen, lactate levels increase with brain activity (12, 13). During seizures, when neurons are maximally activated, lactate levels exceed 10 mM (13, 14) despite substantial *elevations* in blood oxygen.

Presently we understand that lactate can serve a useful purpose by shuttling energy from locations where it is synthesized, like white fibers in skeletal muscle or astroglia in brain, to other locations, like red muscle fibers or pre- and post-synaptic neurons, where it can be oxidized (2). Its residual reductive capacity to provide energy is not wasted by the body and its formation is not caused by a limited supply of oxygen. Muscle and brain do not move away from normal physiology when starting to work, as the earlier paradigm had long accepted. Obviously lactate is not simply an unwanted by-product, but rather is purposefully synthesized during work in order to meet a normal physiological need. However, the nature of that normal physiological mechanism has not been found and until the previous hypothesis that postulated a deficiency of oxygen has been discarded, the reason for lactate generation during muscle and brain work will remain unknown. In this paper, we coordinate results from several avenues of recent research on the energetics of skeletal muscle and brain to propose a model for lactate production during muscle work under well-oxygenated conditions and a rationale for this mechanism. The evidence for the model in muscle is then reviewed, primarily for muscle but also for brain.

9.1. MODEL FOR NON-OXIDATIVE PRODUCTION OF LACTATE UNDER CONDITIONS IN EXERCISING MUSCLE

The energetic mechanisms that have been proposed for muscle and brain reflect experimental measurements of fuels and oxygen consumption made on laboratory times of seconds to hours. However neuronal spiking and muscle contractions occur in milliseconds and energy is seen to be supplied on this time scale in support of these rapid processes, as shown by the ^{31}P NMR experiments of Chung *et al.* (15), discussed above. To understand how these rapid processes are fueled, we must consider the rate of energy delivery. Energy from ATP hydrolysis is produced much faster anaerobically (i.e. by glycolysis or glycogenolysis converting glucose to lactate) than non-oxidatively, in which lactate is ultimately oxidized to CO_2 (16). Thus, with the onset of work we propose that glycolysis and glycogenolysis produce lactate to meet, along with phosphocreatine (PCr; the traditional temporal energy buffer), the high demand for ATP. During the intercontraction period a fraction of the lactate is oxidized to restore PCr and glycogen stores. The remaining lactate is then transported out of the muscle from its place of origin into the blood and thence to other locations where it can be eventually oxidized or converted in the liver to glucose. Because transport is concentration-dependent, the concentration of lactate in a regularly contracting muscle will increase until a steady-state balance is reached between transport out of the muscle cell and net lactate production via this mechanism. Hence, the driving force for the shuttling of lactate is its high concentration, created by the need for rapid non-oxidative energy production that exceeds the need of the muscle cell for lactate oxidation.

When measurements are made in a particular organ, like muscle or brain, most often they are made in timescales of seconds to hours. At these timescales it will appear that more glucose or glycogen has been consumed than expected from the oxygen consumption, leading to the conclusion that the work is supported by the inefficient anaerobic production of energy. However the inefficiency is not as severe as has been thought, because from the viewpoint of the body the lactate is either oxidized (in the muscle or other organs, including other muscle regions as proposed by Brooke) or reconverted to glucose. When glycogen is used as the main source of lactate, as in the muscle, the glycogen is partially to fully (see below) restored during the intercontraction period when most of the oxidative ATP production takes place.

9.2. EVIDENCE FOR RAPID ENERGY CONSUMPTION IN MUSCLE

The conventional view of short-term energetics in the muscle is that PCr supplies almost all of the energy needed for a sustained burst of contractions lasting less than 10 s, after which it is replaced by glycogenolysis. Although this view is presented in textbooks, it is not supported by experiments. In a recent review, Greenhaff and Timmons (17) report, ‘It is now accepted, however, that PCr hydrolysis and lactate production do not occur in isolation, and that both are initiated rather rapidly at the onset of contraction.’

In moderate-intensity exercise, the contractile period can be short, and sequential contractions will generally be separated by durations several times longer than the contractile time. Just how rapidly energy is obtained from PCr during single contractions was shown directly by the brilliant experiments of Chung *et al.* (15). Reasoning that even the fastest freeze-clamp measurements, with a time resolution of ~100 ms, did not follow high-energy phosphates during contraction and relaxation, they developed a gated ^{31}P NMR technique with ~1 ms time resolution to measure phosphate metabolites in the rat gastrocnemius muscle during 1 Hz stimulation. ATP remained constant at all times during contractions, whereas PCr immediately fell by 3 μM g tissue^{-1} twitch^{-1} with a half-time of 8 ms. PCr then recovered with a half-time of 14 ms so that the whole PCr response to a twitch was finished in ~40 ms, too rapidly to have been observed by even the fastest extraction methods (see Figure 8.5). Chung *et al.* showed that this consumption of PCr/twitch is ~40 times greater than the values reported by dividing the drop in PCr after minutes of contraction by the number of twitches. This ^{31}P NMR study showed that ‘the traditional method of calculating PCr/twitch underestimates the high energy consumption arising from the drop and subsequent restoration of the PCr pool during the millisecond contraction cycles’ (15).

9.3. THE INTERPLAY BETWEEN GLYCOGEN, PCr AND LACTATE OXIDATION TO PRODUCE ATP DURING AND AFTER A MUSCLE CONTRACTION

Chung’s experiments show that PCr cannot be the ultimate source of energy in contracting muscle. At a cost of 3 mM PCr/twitch, the muscle would be rapidly depleted of energy unless PCr were being replenished. The experiments show that, in fact, PCr is recovering between contractions. The question then becomes not whether PCr is supporting the contractions but what source of energy is replenishing the PCr levels following contraction.

The rapid resynthesis of PCr during a millisecond contractile period would necessarily utilize non-oxidative means – in skeletal muscle the most likely candidate is ATP production by glycogenolysis. In working muscle, the ATP required during relaxation could not come from glycolytic intermediates because the combined basal concentration of intramuscular glucose and glucose-6-phosphate is below 0.3 mM, which is insufficient to provide enough flux even if completely depleted (18, 19). Furthermore, glucose transport flux, which is effectively unidirectional and therefore equal to the average net uptake measured during exercise, is at least 100 times too low to support the millisecond flux (20, 21). We suggest that glycogenolytic ATP production supplies the energy for millisecond bursts. This hypothesis implies that $\sim 1\,\mu\text{M}\,\text{g tissue}^{-1}$ of glycogen subunits is consumed during each contraction to refill the ATP and PCr pools. However, since glycogen concentration is not depleted even after several dozen contractions, some resynthesis of glycogen must occur between twitches. Without the extreme time pressure of contractions, the ATP required for this resynthesis can be achieved via somewhat slower oxidative means.

Although oxidation cannot supply ATP in milliseconds it could resynthesize glycogen in the ~1 s interval between even rapid contractions. Measured rates of oxygen consumption vary, seemingly dependent upon the measurement method. No reports reach the rate of $1\,\mu\text{M}\,\text{g tissue}^{-1}\,\text{twitch}^{-1}$, but the fastest values reported (16) do reach several $\mu\text{M}\,\text{g tissue}^{-1}\,\text{s}^{-1}$. Hence it is possible that the glycogen pools, which are decreased to refill PCr and ATP in milliseconds, are refilled by the energy supplied oxidatively in the longer period, no shorter than ~1 s, between contractions. In this model, glycogen will decrease non-oxidatively in milliseconds to refill the ATP/PCr pool while it will be resynthesized oxidatively in periods of ~1 s between contractions.

Watchko *et al.* (22) created mutant mouse strains missing either or both forms of creatine kinase and compared functional diaphragm muscle performance with controls during isotonic activation. Both singly mutated forms showed no difference from controls in force, velocity, power, time of sustained shortening and work output, whereas in the doubly mutated animal CK (−/−), these were somewhat reduced. In CK (−/−) animals the velocity of shortening decreased by 20 %, maximum power decreased by 16 %, work decreased by 30 %, and the time to sustain shortening during repetitive isotonic activation decreased by 40 %. In contrast to these moderate functional decreases, the creatine kinase activity in CK (−/−) mutants had decreased to less than 1 % of control values.

Their results showed that PCr did not serve as a steady-state source of energy. Even with only 1 % of creatine kinase activity, the muscle's force was only partially (~20 %) reduced (22). These experiments force us to discard the textbook notion that PCr is necessary to provide the short-term energy for contraction. They do not deny a role for PCr in providing the proximate energy for contraction by buffering ATP, but argue that energetic pathways supply the ultimate energy for contraction.

9.4. EVIDENCE FOR GLYCOGEN'S ROLE IN GENERATING ATP DURING CONTRACTION

The proposal that glycogenolysis is the energy source for exercise is supported by studies of patients with McArdle's disease. In this genetic metabolic disorder, glycogen phosphorylase is completely inactive. Patients suffer from exercise intolerance and muscle cramping. ^{31}P NMR studies (23) of the resting muscle in these patients showed normal PCr, ATP and P_i levels but a decidedly higher pH (~7.2) than in controls (7.02 ± 0.01). During either moderate or high intensity–short term exercise, PCr decreased much more rapidly than in controls, whereas the pH remained high, and ATP levels remained constant. In control subjects exercising at higher intensity, an equivalent drop in PCr was shown to be accompanied by a drop in pH, which is indicative of net lactate production. The inferred minimal lactate production in the McArdle's patients relative to control subjects is evidence that muscle must use glycogen to satisfy its contractile energy requirements. Furthermore, the normal levels of PCr present in these patients did not supply enough energy to support mild exercise in the absence of glycogenolytic flux.

9.5. BIOCHEMICAL EVIDENCE FOR THE MODEL

The proposed model is consistent with biochemical data on the activities of glycogen phosphorylase, glycogen synthase and glucose transport. First, numerous reports find glycogen phosphorylase in its active form under conditions where glycogen concentrations are constant over the long timescales examined (24, 25). On the basis of these findings, and in view of the rapid glycogen turnover proposed here (and experimentally supported below), the allosteric explanation of phosphorylase control *in vivo* should be re-evaluated. Second, a large number of *in vitro* and *in vivo* studies have shown that the rate of glycogen synthesis increases as glycogen concentration decreases (26, 27). The lower concentration of glycogen maintained during more intense exercise is consistent with high rates of synthesis only if there is an increased degradation of glycogen during the intense exercise. In this respect, the dynamic glycogen concentration explains these observations – far more glycogen is being consumed than is apparent from the steady-state observation and maintaining a lower glycogen concentration allows this consumption/restoration cycle to proceed more rapidly during the more intense work.

Finally, the glucose transporter activity increases with exercise, consistent with the hypothesis that increased glycogen turnover takes place despite maintenance of constant glycogen concentrations.

9.6. EVIDENCE FROM ^{13}C NMR OF HUMAN MUSCLE FOR GLYCOGEN TURNOVER DURING EXERCISE

There are uncertainties about the metabolic roles of muscle glycogen in exercise. High glycogen levels improve endurance (28), whereas depletion of glycogen is often associated with the onset of fatigue (29). However, the specific need for glycogen, especially in the presence of sufficient blood glucose, has not been explained, and no specific connection between glycogen levels and fatigue has been possible (30). Nor have the steady-state measurements of glycogen concentrations been related to rapid energy bursts during the millisecond contractions of muscle fibers. Holloszy and Kohrt (31), in summarizing fundamental questions that remain about muscle energetics, led off with 'Why is muscle glycogen necessary for exercise of moderate and high intensities?'

We suggest that ^{13}C NMR experiments of glycogen during exercise have provided a potential answer to this question. Many experiments have shown that the glycogen C-1 resonance is a narrow, well-resolved line in the ^{13}C NMR spectrum and that in muscle and liver the 1.1 % natural abundance ^{13}C NMR peak has good signal strength (see Figure 9.1). Comparison with calibrated solutions and with muscle biopsy data showed that the ^{13}C peak comes from all of the glucose moieties in high-molecular-weight glycogen. Thus, its intensity can be used to accurately assess the concentration of glycogen. Since the accuracy of this determination is several times better than that provided by biopsy measures (see Chapters 2 and 6), it has been possible to extend some of the earlier studies. One relevant illustration of the improvement over earlier work is seen in Figure 9.2, which plots glycogen concentrations during medium intensity plantar flexion exercise of the human gastrocnemius muscle over several hours (32). Although earlier biopsy measurements had supported the suggestion that glycogen reached constant levels during this exercise, the high degree of constancy and its dependence upon exercise level in individuals was certified by the NMR results. The constant level of glycogen depended upon the degree of exercise, reaching lower levels

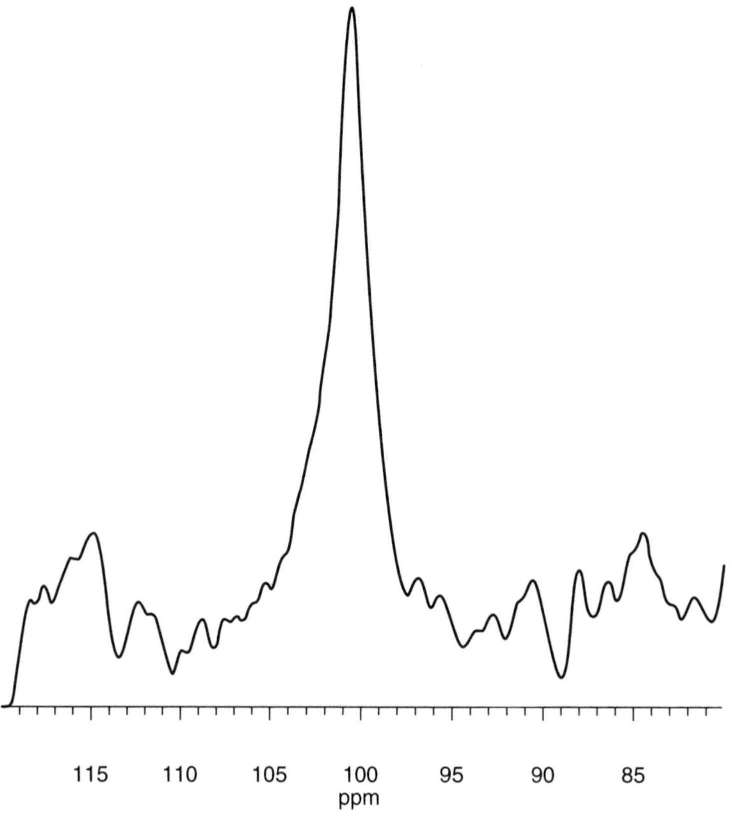

Figure 9.1. Glycogen C-1 resonance from human gastrocnemius at 100.5 ppm obtained by natural abundance ^{13}C nuclear magnetic resonance spectroscopy at 4.7 T. The 5 min scan was performed while the subject was at rest. Signal-to-noise ratio is 21:1.

with increasing workloads. This level depended upon the load, but not upon previous history, as shown in Figure 9.3. For example, following an exercise course of intense toe lifts that resulted in a ~60 % depletion, recovery was biphasic (see Chapter 6). The initial recovery was rapid and independent of insulin while the remainder of the recovery to basal levels was slower and insulin-dependent. However, as shown in Figure 9.3, if medium intensity exercise of the same muscle was conducted after the intense depletion, glycogen levels did not return to the pre-exercise baseline, but to the value characteristic of the plantar flexion exercise. The ability of the muscle either to synthesize or deplete the glycogen stores highlighted the question: does the constant level of glycogen observed reflect a balanced turnover of synthesis and degradation or simply an inert pool? This question was answered (Figure 9.4) by an infusion of labeled 1-^{13}C glucose that was incorporated at a substantial rate into the glycogen pool. Thus, during the several minutes needed for the NMR measurement the glycogen was in a dynamic state of synthesis leading to glucose uptake and degradation leading to lactate accumulation (32).

9.7. LACTATE EFFLUX IN BRAIN

Because of its isolated blood flow system, it has been possible to measure glucose and oxygen uptake by specific brain volumes. Modern brain imaging methods during stimulation have correlated lactate generation

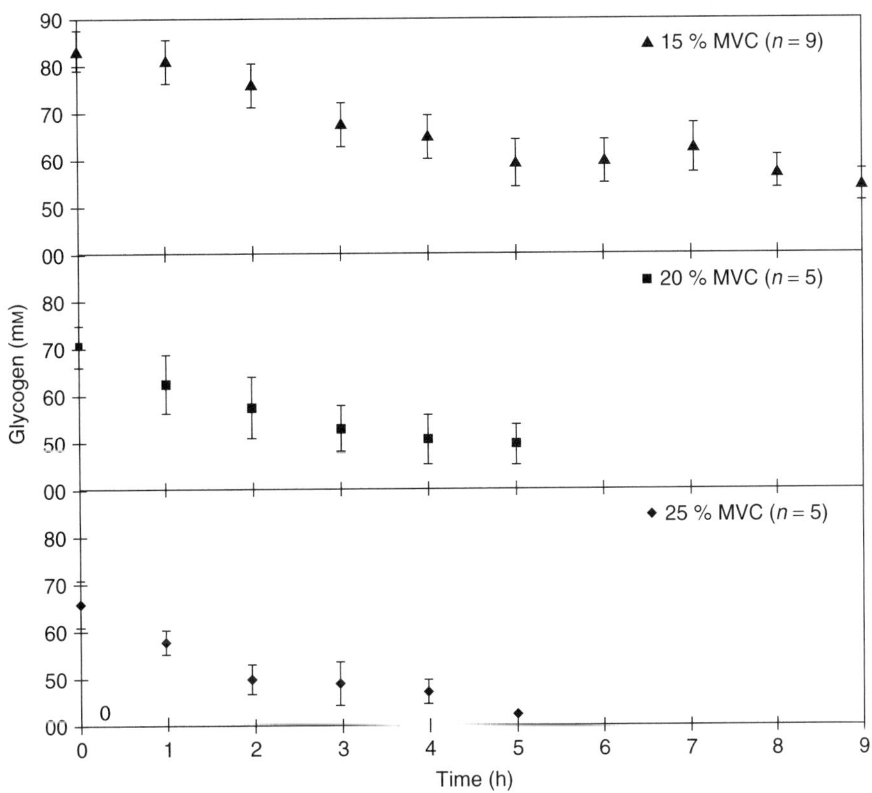

Figure 9.2. Gastrocnemius glycogen concentration vs exercise time for low-intensity exercise from rest at 15, 20 and 25 % of maximal voluntary contraction (MVC). Values are means ± SE.

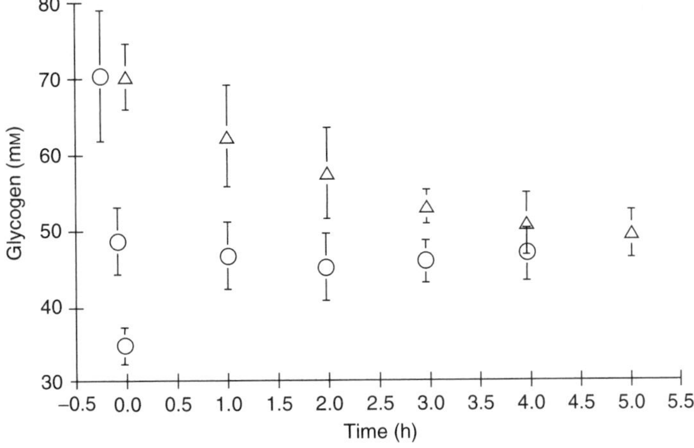

Figure 9.3. Gastrocnemius glycogen concentration vs time for low-intensity exercise (20 % MVC) from rest (Δ, $n = 5$) and for low-intensity exercise (20 % MVC) following predepletion exercise (○, $n = 5$). Values are means ± SE.

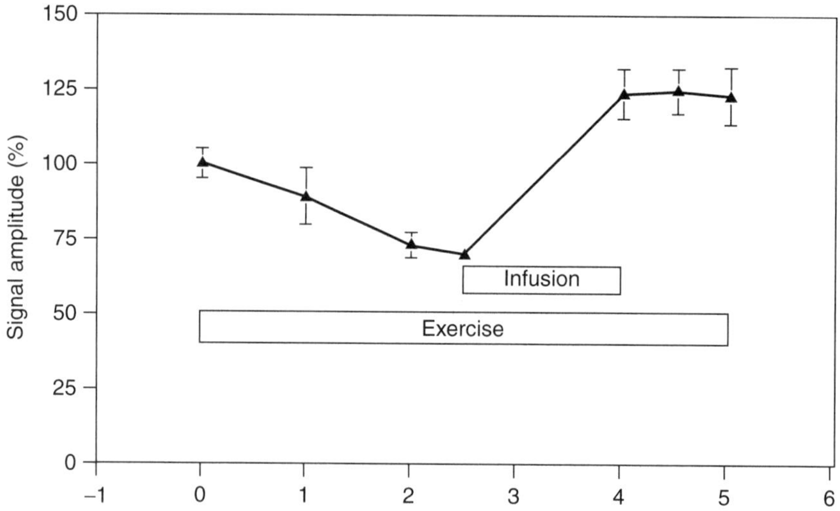

Figure 9.4. Change in gastrocnemius glycogen 1-^{13}C signal amplitude with exercise before and after infusion of 99 % enriched 1-^{13}C glucose. Values are means ± SE for five subjects. Boxes denote exercise and infusion periods. Following infusion, the mean signal amplitude rose to 123 ± 7 % of mean resting amplitude, while plasma enriched in 1-^{13}C glucose rose to 21 ± 2 % above natural abundance.

with glucose and oxygen uptake. Brain lactate levels, like those of muscle, rise during increased functional activity. A surprising result of cerebral energetic studies has been the following: during activation the oxygen–glucose consumption ratio falls away from the approximately 6:1 value of complete oxidation found at the awake resting state to an extreme of 4:1 during the most intense activations observed, seizures (13, 33). Simultaneously, steady-state lactate levels increase from 0.5 mM at rest to ~10 mM during seizures. Intermediate intensity activations, such as those found in visual cortex during exposure to flashing lights, cause lactate to increase slightly above rest values (12). As in the proposed muscle model, the lactate generated during activations can either be consumed later by the producing cell, or moved to other regions via lactate transporters. It was expected (in accordance with the model for lactate production proposed for muscle) that the extra glucose uptake, after conversion to lactate, would be effluxed by the lactate transporters. The available brain data were shown to be compatible with this model, although the scarcity of data, and the uncertain range of lactate transporter kinetic parameters were not able to provide really strong support for the model (33). Nevertheless, there is reason to believe that the steady-state muscle mechanism, in which lactate concentration and its efflux are a measure of the continuing excess rapid, non-oxidative glycolytic flux, are also applicable to brain. In muscle the glycolytic intermediates, including intracellular glucose, are not concentrated enough to supply the rapid millisecond energy, which led to the proposal that the concentrated glycogen supplied the energy.

9.8. PROPOSED ROLE FOR GLYCOGEN DURING FATIGUE

A long-standing question in studies of exercise physiology is why, under conditions where there is adequate fuel from other sources (glucose, fatty acids) to sustain function, the depletion of glycogen leads to fatigue. The model proposed and results described here may provide a solution to this paradox. Glycogen consumption has traditionally been linked to fatigue as the source of excess lactate and associated acidification, while glycogen depletion has been equated to an inability to provide sufficient energy to continue work. However, these explanations are incomplete. In fatigued McArdle patients, phosphorylase deficiency results in higher pH than in resting controls, and many other experiments have dissociated lactate generation (34) and acidosis (35) from muscle fatigue. While it may play a role, particularly in intense anaerobic exercise, lactate acidosis is not the exclusive cause of fatigue.

Normal subjects performing moderately heavy exercise will experience fatigue at low levels of glycogen – even though the net usage of glycogen for energy is insignificant (Figure 9.2). In one study, glycogen levels were measured during exercise with and without continuous ingestion of glucose (30). Glycogen decreased at about the same rate in both cohorts, but because the feeding subjects started at higher concentrations they took 1 h longer to fatigue. In both cases fatigue occurred at a finite, apparently constant, concentration of glycogen, when glucose consumption had almost completely replaced net glycogen consumption as an energy source. The paradoxical findings that the net energy supplied by glycogen at the fatigue point of muscle is small relative to the total energy production can be explained by the glycogen shunt model. Fatigue may be the result of an inability to provide the very high initial power requirements of the contraction through glycogenolysis. This would happen if the glycogen levels dropped to zero momentarily, but also if at low concentrations some of the glycogen particles were reduced to near their glycogenin core, which might cause fluctuations in the binding sites for glycogen synthase and glycogen phosphorylase.

9.9. SUMMARY OF THE GLYCOGEN SHUNT AND LACTATE GENERATION

(1) Support of muscle contraction requires rapid non-oxidative ATP production on the millisecond timescale.

(2) Because of the limited supply of glycolytic intermediates in muscle (including intramuscular glucose) and the ability of glycogen phosphorylase to rapidly increase its activity, glycogenolysis is proposed to make the major contribution to the energy requirements during the milliseconds of contractions.

(3) Between contractions the ATP required for resynthesizing the glycogen, PCr and repumping ion gradients comes mainly from mitochondrial oxidation of lactate and lipids.

(4) Because only a fraction of the lactate produced needs to be oxidized to restore these energy pools, lactate will accumulate in the muscle cell until transport out of the cell equals net production. This accumulation does not reflect the need for inefficient anaerobic glycolysis to produce net power, but rather a mismatch between the efficiency of short-term ATP production (via glycolysis and glycogenolysis) and the oxidative processes which provide most of the net energy production over a contraction–intercontraction cycle.

(5) Fatigue occurs during heavy aerobic exercise where the glycogen concentrations are low but, to a first approximation, constant. In the proposed mechanism of muscle fatigue, fluctuations in glycogen availability result in insufficient ATP to support the rapid contraction periods.

(6) Since refilling the glycogen and PCr pools and repumping of ion gradients post-contraction depend upon oxidative sources of ATP, down-regulated oxidative processes, like mitochondrial myopathies, could cause fatigue. Here the oxidative energy refilling energy sources is inadequate.

(7) The model shows that lactate is not an undesirable byproduct of inefficient anaerobic glycolysis but rather a critical temporal buffer to meet the power demands of muscle contraction. It also provides an

explanation for how fatigue can occur in the presence of low, but still measurable glycogen levels, previously a mystery.

REFERENCES

1. Brooks GA (1985) *Circulation, Respiration and Metabolism: Current Comparative Approaches*, ed. Gillies R. Springer: Berlin; 208–218.
2. Brooks GA (2001) Lactate shuttles in nature. *Biochem. Soc. Trans.* **30**, 258–264.
3. Pellerin L and Magistretti P (1994) Glutamate uptake into astrocytes stimulates aerobic glycolysis: a mechanism coupling neuronal activity to glucose utilization. *Proc. Natl Acad. Sci. USA* **91**, 10625–10629.
4. Magistretti P, Pellerin L, Rothman DL and Shulman RG (1999) Perspective: neuroscience 'energy on demand'. *Science* **283**, 496–497.
5. Brooks GA, Butterfield GE, Wolfe RR, Groves BM, Mazzeo RS, Sutton JR, Wolfel EE and Reeves JT (1991) Decreased reliance on lactate during exercise after acclimatization. *J. Appl. Physiol.* **71**, 333–341.
6. Stanley WC, Gertz EZ, Wisneski JA, Morris DL, Neese R and Brooks GA (1986) Lactate extraction during net lactate release in legs of humans during exercise. *J. Appl. Physiol.* **60**, 1116–1120.
7. Jöbsis FF and Stainsby WN (1968) Oxidation of NADH during concentrations of circulated mammalian skeletal muscle. *Respir. Physiol.* **4**, 292–300.
8. Richter EA, Kiens B, Saltin B, Christensen NJ and Savard G (1988) Skeletal muscle glucose uptake during dynamic exercise in humans: role of muscle mass. *Am. J. Physiol.* **254**, E555–E561.
9. Connett RJ, Gayeski TEJ and Honig CR (1984) Lactate accumulation in fully aerobic, working dog gracilis muscle. *Am. J. Physiol.* **246**, H120–H128.
10. Mole PA, Chung Y, Tran TK, Sailasuta N, Hurd R and Jue T (1999) Myoglobin desaturation with exercise intensity in human gastrocnemius muscle. *Am. J. Physiol.* **277**, R173–R180.
11. Ogawa S, Tank DW, Menon R, Ellermann JM, Kim SG, Merkle H and Ugurbil K (1992) Intrinsic signal changes accompanying sensory stimulation: functional brain mapping with magnetic resonance imaging. *Proc. Natl Acad. Sci. USA* **89**, 5951–5955.
12. Prichard JW, Rothman DL, Novotny EJ, Petroff OAC, Kuwabara T, Avison MJ, Howseman A, Hanstock CC and Shulman RG (1991) Lactate rise detected by ^{1}H NMR in human visual cortex during physiologic stimulation. *Proc. Natl Acad. Sci. USA* **88**, 5829–5831.
13. Siesjo B (1978) *Brain Energy Metabolism.* Wiley: New York.
14. Patel AB, Chowdhury GMI, de Graaf RA, Rothman DL, Wang B, and Behar KL (2003) Evidence that GAD65 mediates a major fraction of GABA synthesis during seizures. *11th Scientific Meeting and Exhibition of International Society of Magnetic Resonance in Medicine*, 10–16 July 2003, Toronto, Ontario.
15. Chung Y, Sharman R, Carlsen R, Unger SW, Larson D and Jue T (1998) Metabolic fluctuation during a muscle contraction cycle. *Am. J. Physiol.* **274**, C846–C852.
16. He ZH, Bottinelli R, Pellegrino MA, Fereczi MA and Reggiani C (2000) ATP consumption and efficiency of human single muscle fibers with different myosin isoform composition. *Biophys. J.* **79**, 945–961.
17. Greenhaff PL and Timmons JA (1998) Interaction between aerobic and anaerobic metabolism during intense muscle contraction. *Exercise Sport Sci. Rev.* **26**, 1–30.
18. Shulman RG, Bloch G and Rothman DL (1995) *In vivo* regulation of muscle glycogen synthesis and the control of glycogen synthesis. *Proc. Natl Acad. Sci. USA* **92**(19), 8535–8542.
19. Roussel R, Carlier PG, Robert J-J, Velho G and Bloch G (1998) NMR studies of glucose transport in human skeletal muscle. *Proc. Natl Acad. Sci. USA* **95**, 1313–1318.
20. Klip A and Paquet MR (1990) Glucose transport and glucose transporters in muscle and their metabolic regulation. *Diabetes Care* **13**, 228–243.
21. Goodyear LJ, Hirshman MF, King PA, Horton ED, Thompson CM and Horton ES (1990) Skeletal muscle plasma membrane glucose transport and glucose transporters after exercise. *J. Appl. Physiol.* **68**, 193–198.
22. Watchko JF, Daood MJ, Sieck GC, LaBella JJ, Ameredes BT, Koretsky AP and Wieringa B (1997) Combined myofibrillar and mitochondrial creative kinase deficiency impairs mouse diaphragm isotonic function. *J. Appl. Physiol.* **82**(5), 1416–1423.

23. Ross BD, Radda GK, Gadian DG, Rocker G, Esiri M and Falconer-Smith J (1981) Examination of a case of suspected McArdle's syndrome by ^{31}P nuclear magnetic resonance. *New Engl. J. Med.* **304**(22), 1338–1342.
24. Ren J-M and Hultman E (1990) Regulation of phosphorylase a activity in human skeletal muscle. *J. Appl. Physiol.* **69**(3), 919–923.
25. Ren J-M and Hultman E (1989) Regulation of glycogenolysis in human skeletal muscle. *J. Appl. Physiol.* **67**(6), 2243–2248.
26. Danforth WH (1965) Glycogen synthetase activity in skeletal muscle. Interconversion of two forms and control of glycogen synthesis. *J. Biol. Chem.* **240**, 588–593.
27. Maehlum S and Hermansen L (1978) Muscle glycogen concentration during recovery after prolonged severe exercise in fasting subjects. *Scand. J. Clin. Lab. Invest.* **38**(6), 557–560.
28. Coyle EF, Coggan AR, Hammert MK and Ivy JL (1986) Muscle glycogen utilization during prolonged strenuous exercise when fed carbohydrate. *J. Appl. Physiol.* **61**(1), 165–172.
29. Saltin B and Gollnick PD (1988) *Exercise, Nutrition and Energy Metabolism*, ed. Horton ES and Tarjung RL. Macmillan: New York; 45–71.
30. Hargreaves M (1997) Interactions between muscle glycogen and blood glucose during exercise. *Exercise Sport Sciences Rev.* **25**, 21–39.
31. Holloszy JO and Kohrt WM (1996) Regulation of carbohydrate and fat metabolism during and after exercise. *A. Rev. Nutr.* **16**, 121–138.
32. Price TB, Taylor R, Mason GF, Rothman DL, Shulman GI and Shulman RG (1994) Turnover of human muscle glycogen with low-intensity exercise. *Med Sci Sports Exercise* **26**, 983–991.
33. Shulman RG, Hyder F and Rothman DL (2001) Lactate efflux and the neuroenergetic basis of brain function. *NMR Biomed* **14**, 389–396.
34. Nielsen OB, DePaoli F and Overgaard K (2001) Protective effects of lactic acid on force production in rat skeletal muscle. *J Physiol* **536**, 161–166.
35. Pan JW, Hamm JR, Hetherington HP, Rothman DL and Shulman RG (1991) Correlation of lactate and pH in human skeletal muscle after exercise by ^{1}H NMR. *J. Magn. Reson. Med.* **20**, 57–65.

10

Futile Cycling in Yeast: How to Control Gluttony in the Midst of Plenty

Jan den Hollander

University of Alabama at Birmingham, center for NMR Research and Development, Birmingham, AL 35294, USA

Robert G. Shulman

Yale University School of Medicine, Department of Diagnostic Radiology, PO Box 208024, New Haven, CT 06520-8024, USA

10.1. INTRODUCTION

The pathways of glucose metabolism in yeast are very similar to the same pathways in muscle. Yet they can respond very differently to cellular needs. A primary environmental difference between microorganisms and muscle cells is the abrupt variation in nutrients, which yeast cells can experience. A sudden inflow of

Metabolomics by In Vivo NMR. Edited by R. G. Shulman and D. L. Rothman
 ISBN: 0-470-84719-0

glucose can change conditions from starvation to plenty. The ways in which yeast accommodate these rapid changes show the versatility of the glucose pathways in different cells. This versatility is most clearly seen when the cells are in transition from one well-adjusted, stationary state to another, brought about by an environmental change that is fast compared with the rate of cellular adaptation (1). During this transitional time yeast survival is faced with conflicting needs. It would be valuable to take advantage of the rich diet offered by the glucose, yet, at the same time the cells are not adapted to use all the energy that full oxidation of this new, rich fuel would offer. We have studied this transition by classical biochemical methods and by *in vivo* ^{13}C and ^{31}P NMR (2, 3).

10.2. YEAST: THE THREE FACES OF HECATE

Yeast changes in response in the environment are well characterized. Cells in a rich medium containing glucose, nitrogen sources and phosphate grow rapidly. These cells are considered to be catabolite repressed, and under these conditions the cells are adapted to live by fermentation of sugar, and their enzyme levels are tailored to take advantage of this favorable situation. In catabolite repression levels of enzymes from the tricarboxylic acid cycle, the electron transport chain and the oxidative phosphorylation pathway are reduced, while glycolytic enzyme levels are elevated.

Another situation arises when glucose is no longer available. Yeast cells undergo a diauxic shift after they exhaust their glucose supply, but continue to grow on non-fermentable carbon sources such as ethanol or acetate provided oxygen sources are available. These cells are derepressed, that is they have active respiration while the levels of the glycolytic enzymes are reduced. Cells grown in the absence of glucose actually store carbohydrates, trehalose and glycogen, which they synthesize through gluconeogenesis (4, 5) Hence these cells require the enzymes needed for gluconeogenesis: cytosolic malate dehydrogenase, Fru-1,6-P_2-ase, phosphoenolpyruvate carboxykinase and isocitrate lyase (6). Cells growing in the presence of adequate oxygen on non-fermentable carbon sources reach a well-adapted state. These cells, denoted respiratory competent, or derepressed, have active enzymatic pathways beyond pyruvate dehydrogenase in contrast to cells grown on glucose, denoted catabolite repressed, which are not respiratory competent.

Since the glycolytic enzymes are constitutive, present under all conditions, all cells can take up glucose, although at different rates. However when a derepressed cell accustomed to efficiently oxidizing acetate or ethanol is suddenly offered glucose, widespread adaptations via gene expression and catabolite inactivation will start.

A third situation arises in vegetative cells when nitrogen sources are exhausted: cells stop dividing, but may continue to store carbohydrates. In the presence of acetate they may sporulate, to form ascospores. These spores have low metabolic activity, and are well adapted to survive adverse conditions until better times arrive (Chapter 11).

10.3. PASTEUR EFFECT

Many experiments investigating glucose consumption by different states of yeast have been conducted since Louis Pasteur's time. Because of the diverse approaches, there is confusion about what is actually meant by the 'Pasteur effect' (7). Pasteur reported that the amount of yeast formed from a certain amount of sugar is much larger in the presence of oxygen than in its absence (8). Warburg, however, introduced the term 'Pasteur reaction' to describe the difference between aerobic and anaerobic glucose catabolism (9). Meyerhoff measured a factor 5–7 difference in the aerobic and anaerobic rates of glucose *catabolism* by experiments using the Warburg manometer (10). On the other hand, the difference between aerobic and anaerobic glucose *consumption* by derepressed cells has been found to be no more than about a factor 2 (11, 12). These differences between aerobic and anaerobic glycolysis are only found in respiratory-competent cells;

cells grown under repressed conditions derive metabolic energy from glycolysis and do not change their glycolytic rate going from anaerobic to aerobic conditions. In accord with present usage we have used the term 'Pasteur effect' to describe the difference in net flux down the Embden–Meyerhoff–Parnas (EMP) pathway between aerobic and anaerobic glycolysis.

To study the Pasteur effect we exposed derepressed, respiratory-competent cells grown on non-fermentable substrates to glucose. The cells have to adapt to the sudden availability of glucose, and will change during a transition phase while glycolytic enzymes are synthesized, enzymes needed for gluconeogenesis are inactivated through catabolite inactivation (1) and oxidative phosphorylation is repressed. Our yeast NMR studies were conducted under a well-defined set of conditions. Rates of aerobic and anaerobic glycolysis ('the Pasteur effect') were compared either on acetate-grown, derepressed, respiratory-competent yeast or on cells grown into saturation. Exposure to glucose was brief to minimize changes in enzyme levels.

Yeast provided the original ground for *in vivo* NMR studies of metabolic fluxes (13–16). Now generally called magnetic resonance spectroscopy (MRS), its applications form the basis of this book. These yeast studies are not only of historical interest but also provide explanations of several contemporary questions in yeast metabolism, questions with implications for biochemistry in general. The results to be discussed center on ^{13}C NMR measurements of fluxes from ^{13}C-labeled glucose in derepressed *Saccharomyces cerevisiae*. These flux measurements have been compared with ^{31}P saturation transfer measurements of fluxes in the same parts of the pathway. In the saturation transfer experiments, the nuclear spins at one phosphorous nucleus are 'saturated' by a selective radiofrequency field. This specifically labels those spins, and the flow of the labeled spin population during chemical reactions to other sites is detected by the transferred loss of signal (17). The comparison between the rates measured by these two methods has overdetermined metabolic fluxes in the living yeast, and the values obtained by ^{31}P saturation transfer agree with those measured by ^{13}C NMR under the same conditions (15).

The question posed by the original studies was how yeast regulate the glycolytic flux when oxygen is present, and which metabolic intermediates mediate that regulation. With time this question has been rephrased and forwarded by new results. Presently the interpretation focuses on the first steps of the glycolytic pathway where restraints on the rate of glucose catabolism must exist in order to satisfy, but not overstimulate, the subsequent oxidative steps. This potential overstimulation has been termed the 'Turbo effect' (18) and results showing inhibition exercised by trehalose-6-phosphate (T6P) on hexokinase (HKase) have provided a mechanism to restrain the Turbo mechanism (19). This review shows how the early *in vivo* ^{13}C NMR results measuring mass flows in the early steps of the glycolytic pathway agree with and can extend the understanding of flux control reached by the Turbo approach. The ^{13}C NMR experiments have measured the reverse flow or 'futile cycling' in the pathway, a cycling that works together with T6P inhibition to limit the glucose uptake, and to help the cells accommodate, transiently, to the sudden rich environment.

10.4. ^{13}C NMR OF AEROBIC AND ANAEROBIC GLYCOLYSIS IN YEAST

Figure 10.1 shows a comparison of the consumption of [1-^{13}C]glucose by aerobic and anaerobic yeast cells. [1-^{13}C]glucose, as the substrate, allowed the ^{13}C label to be followed into metabolic pools by ^{13}C NMR, thereby measuring the rate of glucose utilization. Derepressed cells consume glucose 1.6–1.81 times faster anaerobically than aerobically, while glucose repressed cells showed no difference between aerobic and anaerobic glucose consumption (Table 10.1), in agreement with earlier findings (12, 13). Combined ^{13}C NMR experiments and ^{14}C tracer studies were used to measure the appearance of the labels in soluble compounds, the gas phase, and macromolecular fractions during aerobic and anaerobic glycolysis (13). The results of those experiments were used to determine the fluxes in the upper part of the EMP pathway (see Figures 10.2 and 10.3). Metabolic fluxes through the enzymes of the EMP pathway were measured by ^{13}C NMR using ^{13}C-labeled glucose to follow the flow of the ^{13}C label into intermediates and end-products (13, 14), and by ^{31}P saturation transfer experiments (3, 15) for direct measurement of metabolic flows. The ^{13}C NMR studies were supplemented with ^{14}C tracer studies to quantify the amount of ^{14}C label that appeared in the gas-phase as CO_2 and in macromolecules (13) (Table 10.2). ^{13}C NMR was also used to determine how much of the ^{13}C label appeared in the storage compound Trehalose. In addition, ^{31}C and ^{31}P NMR were used to measure the concentrations of glycolytic intermediates both *in vivo* (13, 20, 21) and in extracts (14, 16) (Table 10.3).

^{13}C NMR experiments with [1-^{13}C]glucose or [6-^{13}C]glucose provided detailed information about the metabolic flows by following the scrambling of the ^{13}C label between the C-1 and C-6 of the hexoses. The scrambling of the label was shown in Fru-1,6-P_2 and came from the forward and backward flows through the aldolase and triosephosphate isomerase triangle [for a detailed discussion see den Hollander *et al.* (20)], showing close to equal distribution of label between C-1 and C-6 in Fru-1,6-P_2 [data not shown, see den Hollander *et al.* (14)]. Furthermore, from the appearance of the ^{13}C label in the C-1 position of trehalose from [6-^{13}C]glucose or in the position C-6 from [1-^{13}C]glucose, we were able to determine the backward flow through Fru-1,6-P_2-ase [see den Hollander *et al.* (2, 16) for a more detailed discussion]. In acetate-grown cells trehalose was generated and its label scrambling observed only under aerobic conditions (Figure 10.4).

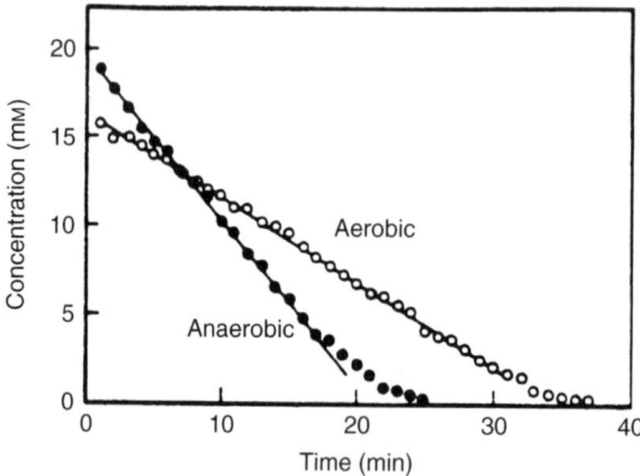

Figure 10.1. Total [1-^{13}C]glucose concentration as a function of time as determined from ^{13}C NMR spectra. Glucose consumption was followed in parallel aerobic and anaerobic experiments with a suspension of acetate-grown yeast cells (20 % wet weight). To determine the total concentration of [1-^{13}C]glucose, the NMR signals from α and β [1-^{13}C]glucose were added (13). Glucose consumption is a factor 1.81 faster anaerobically than aerobically. (Reproduced from den Hollander JA, Ugurbil K, Brown TR, Bednar M, Redfield C and Shulman RG (1986) Biochemistry **25**(1): 203–211 Permission of the American Chemical Society.).

Table 10.1. Effect of oxygen on the maximum glucose utilization rate as measured by ^{13}C NMR

Carbon source used for growth	$V_{max}(-O_2)/V_{max}(+O_2)$
Glucose (repressed)[a]	1.0
Glucose (derepressed)[b]	1.75
Acetate	1.81

[a] Cells were grown on glucose and were harvested at ~25 % of the cell density attained by cultures grown fully to saturation.
[b] Cells were harvested after they were grown to full saturation (13).

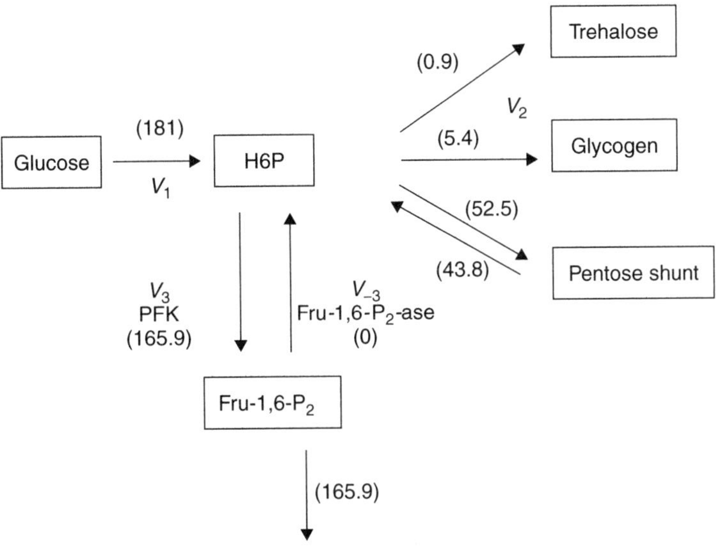

Figure 10.2. Flow diagram for anaerobic glycolysis in acetate-grown yeast. Glucose consumption is scaled compared with Figure 10.3 to take into account the factor 1.81 higher glucose consumption rate measured for anaerobic cells. The numbers in parentheses are the relative fluxes, normalized to a glucose consumption rate of 181, to reflect the factor 1.81 higher anaerobic rate. H6P stands for the combined G6P and Fru-6-P pool.

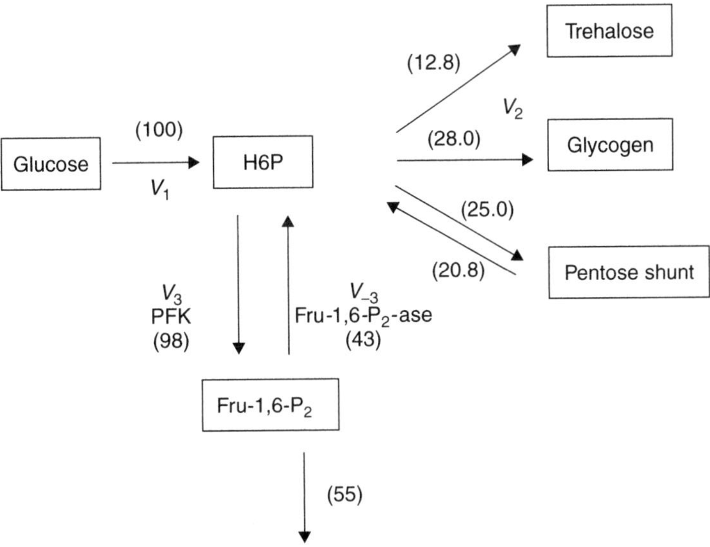

Figure 10.3. Flow diagram of relative flows through the upper part of the glycolytic pathway, for aerobic glycolysis in acetate grown yeast cells. The relative flows were derived from the label distributions tabulated in Table 10.2. It is assumed that most of the labeled CO_2 is formed through pentose shunt activity, and that the pentose phosphates are recycled to the hexose phosphate pool. Glycogen is estimated from the macromolecular pool. The back flow through Fru-1,6-P_2-ase is calculated from the scrambling of the ^{13}C label from C1 to C6 in trehalose.

Table 10.2. Distribution of label from [1-^{13}C] and [1-^{14}C]glucose between different metabolic pools during anaerobic and aerobic glycolysis in acetate-grown *S. cerevisiae*

	Acetate grown		Glucose grown (saturated)	
	Anaerobic	Aerobic	Anaerobic	Aerobic
Trehalose	0.5	12.8	5.0	12.5
Glycerol	25.5	–	12.6	8.7
Ethanol	42.4	29.0	59.6	35.3
Glutamate	0.8	5.9	3.6	22.8
(Other)	1.6	–	5.5	3.4
Gas phase (mostly CO_2)	29.0	25.0	8.0	12.5
Macromolecules (mostly glycogen)	3.0	28.0	5.5	5.0
Totals	102.8	100.7	99.8	100.2

Reproduced from Reibstein D, den Hollander JA, Pilkis SJ and Shulman RG, *Biochemistry* **25**(1): 219–227 by permission of the American Chemical Society.

Table 10.3. Concentrations of intermediates (mM) measured in glycolyzing cells[a]

	Glucose grown (repressed)		Glucose grown (derepressed)		Acetate grown	
	Anaerobic	Aerobic	Anaerobic	Aerobic	Anaerobic	Aerobic
G6P[a]	1.8	1.6	1.0	4.1	0.8	6.9
Fru-6-P[b]	0.6	0.53	0.33	1.37	0.27	2.3
αGP[a]	0.7	0.3	3.2	1.8	5.7	2.0
3PGA[a]	0.4	0.4	<0.2	0.7	0.3	0.8
Fru-1,6-P_2[a]	10.1	6.2	3.1	4.2	1.9	2.2
NDP[a]	0.5	0.4	0.9	0.4	1.1	0.6
NTP[a]	3.7	3.4	3.7	4.8	3.2	4.7
P_i[c]	5	7	10.0	4.0	17	<2
Fru-2,6-P_2[d]			3.2 μM	0.6 μM		
pH[b]	7.25	7.28	7.23	7.52	7.13	7.58

[a] Estimated intracellular concentrations of the levels of intermediates, determined from the ^{31}P NMR spectra of cell extracts, in millimolar intracellular concentrations (14).
[b] Derived from G6P by assuming [Fru-6-P] ≈ $\frac{1}{3}$ [G6P].
[c] Determined from *in vivo* ^{31}P NMR measurements (21).
[d] From enzyme assays (16).

The net glycolytic flow was determined by subtracting the rate of glucose utilization by non-glycolytic pathways (V_2) (i.e. trehalose, and glycogen storage, and net flow into the pentose shunt) from the overall glucose utilization rate (V_1) (Figure 10.3). To obtain the forward flow through PFK this net glycolytic flux was then corrected for backward flux through Fru-1,6-P_2-ase, i.e. V_{-3}, which was measured as described below. Saturation transfer measurements of the unidirectional flow through PFK (V_3), confirmed and validated the overall flux determination (15). It is important to note that the two independent rates V_{-3} and V_3 were overdetermined because three measurements were made, and the results were all consistent (3, 15).

The overall glucose utilization rate was determined by measuring the time course of [1-^{13}C]glucose consumption using ^{13}C NMR (Figure 10.1 and Table 10.1). The rates of non-glycolytic flows such as trehalose and glycogen biosynthesis and the phosphogluconate pathway, which add up to V_2, were evaluated using ^{13}C and ^{14}C-labeled glucose. The ^{13}C-labeled glucose was used to determine amounts of labeled soluble

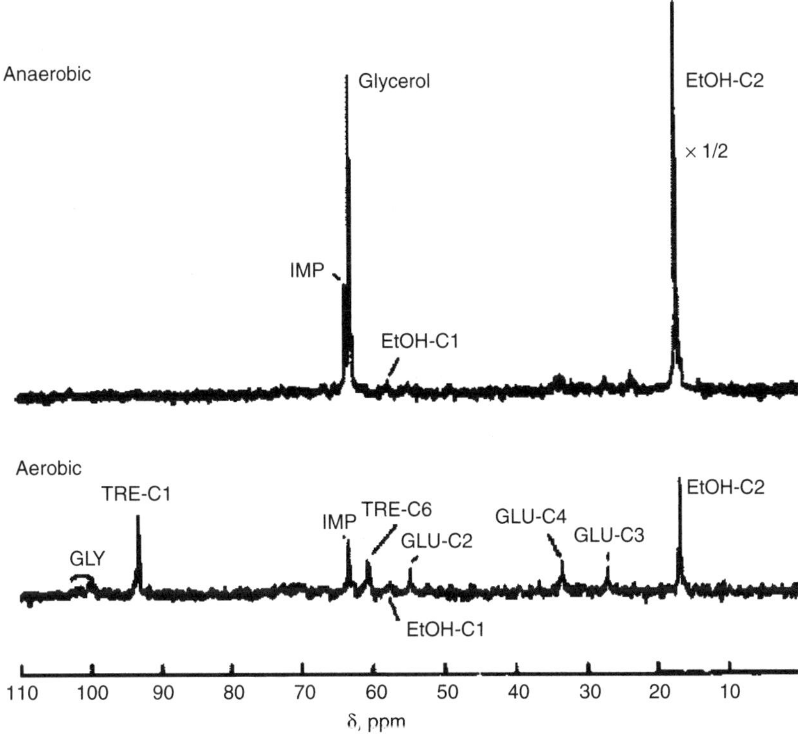

Figure 10.4. ^{13}C NMR spectra obtained after the anaerobic (upper trace) and aerobic (lower trace) utilization of [1-^{13}C]glucose by a suspension of acetate-grown yeast cells. Cell density was 20% wet weight. Note the formation of trehalose in the aerobic condition, with the scrambling of the ^{13}C label from the C_1 to C_6 position, and its absence anaerobically. The relative labeling of the C_6 position of trehalose allows us to calculate the backward flux through Fru-1,6-P_2-ase (13). TRE, trehalose; EtOH, ethanol; GLU, glutamate; and GLY, glycogen. (Reproduced from den Hollander JA, Ugurbil K, Brown TR, Bednar M, Redfield C and Shulman RG (1986) Biochemistry **25**(1): 203–211 Permission of the American Chemical Society.).

end products such as ethanol, glycerol and trehalose. The radioisotope ^{14}C data allowed determination of the amounts of the glucose incorporated into macromolecules (Table 10.2). In similar measurements, the trichloracetic acid-precipitable macromolecular pool had been shown to consist predominantly of polysaccharides (12). Flow through the pentose shunt was evaluated by measuring the appearance of the ^{14}C tracer in the gas phase as CO_2. Hence, it was possible to obtain the difference ($V_1 - V_2$), i.e. the net flow through the EMP pathway.

Upon addition of either [1-^{13}C]glucose or [6-^{13}C]glucose to the cells ^{13}C NMR measurements of the label distribution in trehalose were used to measure Fru-1,6-P_2-ase flow (Figure 10.4). In acetate-grown yeast, Fru-1,6-P_2 was almost equally labeled at C-1 and C-6 during aerobic glycolysis of [6-^{13}C] or [1-^{13}C]glucose. We measured the relative flow through Fru-1,6-P_2-ase (V_{-3}) by the amount of C-1 labeling observed in trehalose from [6-^{13}C]glucose, since all of the C-1 labeling had to come from Fru-1,6-P_2. The relative amounts of labeling at C-1 and C-6 carbons of trehalose were used to determine V_{-3}/V_3, [see den Hollander *et al.* (2, 13) for a detailed discussion of label scrambling in trehalose, and how this is used to determine the relative flow through Fru-1,6-P_2-ase]. It is interesting to note that futile cycling in acetate grown cells was extensive [$V_{-3} \approx 44\%\ V_3$ (13)] under aerobic conditions but not observed under anaerobic conditions (Figure 10.4). The *in vivo* rate of the unidirectional flow through phosphofructokinase (PFK) (V_3) was determined from the equation $V_3 - V_{-3} = (1 - 0.44) \times V_3 = V_1 - V_2$, since both the glucose consumption rate (V_1), and the total flow into storage carbohydrates and pentose shunt (V_2) were measured (Figure 10.4, Table 10.2).

10.5. MANAGING THE METABOLIC PREDICAMENT: TREHALOSE PATHWAY AND FUTILE CYCLING

Only under transient conditions do yeast cells exhibit a small difference in the rate of aerobic and anaerobic glucose consumption. Derepressed cells, exposed to glucose undergo a major overhaul of their metabolic machinery on their way to a stationary repressed state and it is during this transition phase that they exhibit this effect. When those cells are exposed to glucose the yeast cells need to turn on PFK and turn off Fru-1,6-P_2-ase to switch from gluconeogenesis to glycolysis quickly if they are to use glucose efficiently without wasting energy.

Our results demonstrate that, during the switchover from respiration to fermentation, from gluconeogenesis to glycolysis, when derepressed yeast are provided with high glucose, they maintain the flux of the EMP pathway in a narrow range. The danger of too much flux in respiratory competent cells under aerobic conditions in the presence of glucose is that the high adenosine triphosphate (ATP) production would enhance the HKase rate, which would produce large amounts of ATP oxidatively. The runaway aspects of this situation have been denoted as 'turbo' (18, 19) so that negative feedback is required particularly since no negative feedback of HKase by G6P has been demonstrated. However, 'turbo' control was recently shown to be enforced by an intermediate in trehalose synthesis, T6P, which is synthesized from glucose-6-phosphate (G6P) and which acts as a negative effector upon HKase (19, 22). The early NMR results which reported trehalose synthesis were consistent with this mechanism, although they had no information about the connection between trehalose synthesis and inhibition of glycolysis since they predated knowledge of T6P control. The ^{13}C NMR experiments measured mass flow in both directions of the glycolytic pathway. The reverse flux through Fru-1,6-P_2-ase reduces the net EMP forward flux. Its contribution to an increased G6P concentration adds to the negative feedback mechanism of hexokinase by T6P described in the 'turbo' paper (18). Under aerobic conditions, the NMR results show substantial 'futile' cycling occurring through the PFK and Fru-1,6-P_2-ase cycle, but not anaerobically (Figures 10.2 and 10.3). This happens despite the well-known catabolite inactivation of Fru-1,6-P_2-ase when derepressed yeast cells are exposed to glucose. However, our results show that during this transition Fru-1,6-P_2-ase remains active, at least when the cells are aerobic for the several minutes' duration of the NMR experiments. Under anaerobic conditions when Fru-1,6-P_2-ase is always repressed there is no sign of 'futile' cycling through Fru-1,6-P_2-ase. Hence, although catabolite inactivation of Fru-1,6-P_2-ase did not occur in aerobic cells during the time of our measurements, it appears to have occurred in similar cells anaerobically maintained in a steady state. Whether this reflects the rates of Fru-1,6-P_2-ase inactivation or another metabolic control mechanism is not known.

This difference in Fru-1,6-P_2-ase activity between aerobic and anaerobic glycolysis by derepressed cells suggests that the ongoing Fru-1,6-P_2-ase activity during aerobic glycolysis and the cycling through the trehalose pathway serves an important purpose: yeast enzymes do not just idly spin their wheels, but employ 'futile cycling' to manage a metabolic crisis in this transition phase. Consider what happens to Fru-1,6-P_2. If Fru-1,6-P_2 synthesis by PFK exceeded the demand for Fru-1,6-P_2 by the lower part of the EMP pathway there would be a continued buildup of Fru-1,6-P_2. This was not observed: the concentration of Fru-1,6-P_2 was nearly the same during aerobic and anaerobic glycolysis (7). We propose that yeast cells maintain this homeostasis of Fru-1,6-P_2 by two mechanisms. The first, as revealed by the ^{13}C NMR is by reverse flow through ongoing Fru-1,6-P_2-ase activity. Under aerobic glycolysis in acetate grown cells, the flow through Fru-1,6-P_2-ase was about 44 % of that through PFK. This reverse flux reduced the total net flow through the PFK/Fru-1,6-P_2-ase module of the glycolytic pathway. In addition the reverse flux wasted some of the excess ATP, and stored glucose as trehalose and glycogen for future energy needs. Trehalose synthesis is a reverse mass flow that makes a significant contribution to reducing the glycolytic flux, thereby cooperating with T6P inhibition of the forward flux through HKase and PFK.

10.6. THE ROLE OF PFK

To understand the role of PFK with and without oxygen, the concentrations of the eight effectors of PFK, including the substrate Fru-6-P, were measured *in vivo* and in extracts (16). For six effectors the levels were determined by NMR, while for one they were derived from the NMR data by calculation and for another, Fru-2,6-P_2, they were determined by chemical means. Results established that the response of PFK to any single effector depends on the concentration of other effectors. Therefore, to establish the physiological importance of each effector, *in vivo* concentrations of all eight effectors of PFK were established *in vitro* in a partially purified PFK preparation. The *in vitro* rates reproduced the measured *in vivo* rates of flow through PFK under both aerobic and anaerobic conditions. Sensitivity of the PFK rate to individual effectors was investigated by varying the concentration of each effector. The change in PFK activity during the Pasteur effect included strong and partially compensating contributions from Fru-2,6-P_2 and the substrate Fru-6-P. Other effectors, in particular [ATP], [AMP], [P_i] and pH all contributed to modulating the PFK activity during the switch from anaerobic to aerobic conditions [see Reibstein *et al.* (16), Table 10.2]. Hence this experiment not only reproduced the *in vivo* fluxes in an *in vitro* experiment, but also identified the controlling effectors that are important during the Pasteur effect.

The concentrations measured of effectors and intermediates (Table 10.2) show small differences in ATP and Fru-1,6-P_2 between aerobic and anaerobic glycolysis. However, Fru-2,6-P_2 is five times lower in the aerobic condition, while G6P is four to eight times higher (and therefore Fru-6-P as implied by rapid phosphoglucose isomerase activity). Fru-2,6-P_2 is a potent stimulator of glycolytic flux that stimulates PFK activity and inhibits Fru-1,6-P_2-ase. In the presence of all known effectors (Figure 10.5) at low Fru-6-P concentrations the activity of PFK *in vitro* is an order of magnitude higher in the anaerobic condition. However, the much higher concentration of Fru-6-P found for the aerobic condition counteracts the difference in PFK activity, with the net result that there is only a modest reduction in PFK flux aerobically (~1.8-fold).

Results of Figure 10.5 suggest that in anaerobic glycolysis PFK activity is so high that the flux is limited by availability of the substrate Fru-6-P. Much of this high PFK activity can be ascribed to the high Fru-2,6-P_2 concentration, an effector which inhibits Fru-1,6-P_2-ase, and thus prevents futile cycling during anaerobic glycolysis. There is only a modest reduction of PFK activity despite the 5-fold lower Fru-2,6-P_2 concentration in the aerobic condition, because of the compensating effect of the increased F6P concentration. However, as mentioned above, the lower Fru-2,6-P_2 concentration introduces little or no inhibition of Fru-1,6-P_2-ase. The combined effect of the reduced flow through PFK and the 44 % backward flux through Fru-1,6-P_2-ase results in a reduction of the net flux through the EMP pathway by a factor of 3 (see flow schemes, Figures 10.2 and 10.3). Presumably, the concentration of Fru-2,6-P_2 remains low enough in the transient state of aerobic glycolysis measured here that PFK and Fru-1,6-P_2-ase are both active.

The increased level of Fru-6-P and G6P under aerobic conditions shown in Table 10.2 up-regulates the synthesis of T6P and trehalose through T6P synthase. This provides two ways to modulate the net flow through the upper part of the EMP pathway. First some of the glucose moieties flow into the trehalose and glycogen pools rather than running down the glycolytic pathway. Second, T6P is now

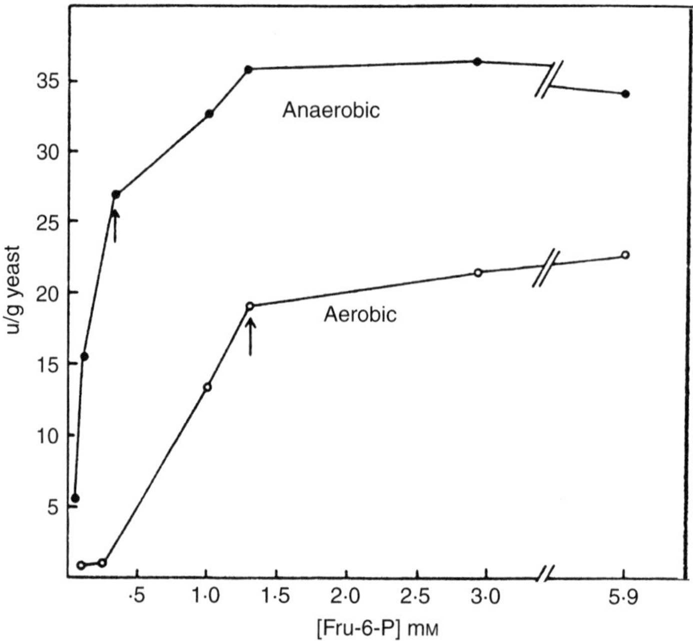

Figure 10.5. Activity of PFK as a function of [Fru-6-P] concentration, with effector concentrations given in Table 10.3, including Fru-6-P but omitting ADP. Arrows show anaerobic and aerobic concentrations of Fru-6-P (Reproduced from Reibstein D, den Hollander J, Pilkis SJ and Shulman RG, Biochemistry 1986; **25**(1): 219–227 by permission of the American Chemical Society.).

known to be a potent inhibitor of HKase (18, 19, 22). Therefore, the trehalose pathway provides a mechanism for down-regulating phosphorylation of glucose, thereby preventing the unlimited buildup of the G6P pool.

The high concentration of HMPs in the aerobic conditions relates the allosteric control of PFK to the futile cycling. Fru-1,6-P_2-ase flux feeds the HMP pool and a high concentration of HMPs increases the rate of trehalose synthesis and of T6P. Increased T6P reduces the HKase glycolytic flux. The high Fru-6-P achieves a 1.6 × reduced PFK flux and at the same time helps trehalose synthesis, reduces HKase activity, and accommodates Fru-1,6-P_2-ase back flux.

We are now able to relate the metabolic fluxes in Figures 10.2 and 10.3 and Table 10.2 with the 1.8-fold reduction of glucose consumption under aerobic conditions to derive the full reduction of flux through the lower part of the EMP pathway. The HKase flux (V_1) matches the uptake and these equal the PFK flux which is reduced ~1.6-fold from the anaerobic condition. However anti-parallel with PFK the upward Fru-1,6-P_2-ase flux (V_{-3}) is ~$0.44V_3$, showing the strong effect of the Fru-1,6-P_2-ase 'futile' cycling. As a result of this reverse flux and the significant, and equal, storage of glucose (V_2), the flux beyond Fru-1,6-P_2 is down by a factor of ~3 compared to the anaerobic flux, in contrast to the glucose uptake which is down by ~1.8-fold. Almost half of the reduction is realized by the back flow. Although it may be coincidental, we note that within experimental error the fluxes through HKase and PFK are equal, so that the fluxes of Fru-1,6-P_2-ase and trehalose synthesis are also equal. Possibly there is a unifying reason for this equality – if so; it evades us at this time.

10.7. CONCLUSION

The results reviewed here on the derepressed cells show that glucose is consumed ~1.8-fold slower in the presence of oxygen. This reduced flux flows through HKase and PFK. The net forward flux is reduced further by back fluxes: (1) from Fru-1,6-P_2 to HMP by the action of Fru-1,6-P_2-ase; and (2) from the HMP pool to trehalose and glycogen. Hence, the flux through the lower part of the glycolytic pathway is further reduced by back flows in the aerobic condition.

The existence of futile cycling seems to depend upon the conditions. In our experiments futile cycling, bypassing both PFK and HKase, does exist in aerobic derepressed cells but not anaerobically. Futile cycling by creating a backflow contributes to solving the problem introduced by the 'turbo' runaway mechanism in that it reduces the oxidative fluxes, by a mechanism that adds to the inhibition of the first steps and also serves to reduce the energy production. Reverse fluxes also store, as trehalose, some of the available glucose not needed for energy. Serving these two obviously valuable functions in the present case, the cycling can hardly be considered 'futile'.

REFERENCES

1. Holzer H (1976) Catabolite inactivation in yeast, *Trends Biochem. Sci.* **1**: 178–181.
2. den Hollander JA and Shulman RG (1983) ^{13}C NMR studies of *in vivo* kinetic rates of metabolic processes, *Tetrahedron* **39**(21): 3529–3538.
3. Campbell-Burk SL and Shulman RG (1987) High-resolution NMR studies of *Saccharomyces cerevisiae*, *A. Rev. Microbiol.* **41**: 595–616.
4. Francois J and Parrou JL (2001) Reserve carbohydrates metabolism in the yeast *Saccharomyces cerevisiae*, *FEMS Microbiol Rev* **25**: 125–145.
5. den Hollander JA, Behar KL and Shulman RG (1981) ^{13}C NMR study of transamination during acetate utilization by *S. cerevisiae*, *Proc. Natl Acad. Sci. USA* **78**: 2693–2697.
6. Entian K-D and Barnett JA (1992) Regulation of sugar utilization by *Saccharomyces cerevisiae*, *Trends Biochem. Sci.* 506–510.
7. Lagunas R (1981) Is *Saccharomyces cerevisiae* a typical facultative anaerobe? *Trends Biochem. Sci.* 201–203.
8. Louis Pasteur Scientific Papers, The Harvard Classics (1909–1914) *The Physiological Theory Of Fermentation I. On the Relations Existing Between Oxygen and Yeast*; www.bartleby.com/38/7/1.html.
9. Warburg O (1926) Über die Wirkung von Blausäureäthylester (Äthylcarbylamin) auf die Pasteursche Reaktion. *Biochem. Z.* **172**: 432–444.
10. Meyerhof O (1925) Über den Einfluß des Sauerstoffs auf die alkoholische Gärung der Hefe. *Biochem. Z.* **162**: 43–86.
11. Lynen F (1961) *Symp. Soc. Gen. Physiol.* **8**: 289.
12. Stickland LH (1956) The Pasteur effect in normal yeast and its inhibition by various agents. *Biochem. J.* **64**: 503–515.
13. den Hollander JA, Ugurbil K, Brown TR, Bednar M, Redfield C and Shulman RG (1986) Studies of anaerobic and aerobic glycolysis in *Saccharomyces cerevisiae*. *Biochemistry* **25**(1): 203–211.
14. den Hollander JA, Ugurbil K and Shulman RG (1986) ^{31}P and ^{13}C NMR studies of intermediates of aerobic and anaerobic glycolysis in *Saccharomyces cerevisiae*. *Biochemistry* **25**(1): 212–219.
15. Campbell-Burk SL, den Hollander JA, Alger JR and Shulman RG (1987) ^{31}P NMR saturation-transfer and ^{13}C NMR kinetic studies of glycolytic regulation during anaerobic and aerobic glycolysis. *Biochemistry* **26**(23): 7493–7500.
16. Reibstein D, den Hollander JA, Pilkis SJ and Shulman RG (1986) Studies on the regulation of yeast phosphofructo-1-kinase: its role in aerobic and anaerobic glycolysis. *Biochemistry* **25**(1): 219–227.

17. Alger JR and Shulman RG (1984). NMR studies of enzymatic rates *in vitro* and *in vivo* by magnetization transfer. *Q. Rev. Biophys.* **17**: 83–124.
18. Teusink B, Walsh MC, van Dam K and Westerhoff HV (1998) *Trends Biochem. Sci.* **23**: 162–169.
19. Blázquez MA, Lagunas R, Gancedo C and Gancedo JM (1993) Trehalose-6-phosphate, a new regulator of yeast glycolysis that inhibits hexokinases. *FEBS Lett.* **239**: 51–54.
20. den Hollander JA, Brown TR, Ugurbil K and Shulman RG. (1979) ^{13}C nuclear magnetic resonance studies of anaerobic glycolysis in suspensions of yeast cells. *Proc. Natl Acad. Sci. USA* **76**(12): 6096–6100.
21. den Hollander JA, Ugurbil K, Brown TR and Shulman RG. (1981) Phosphorus-31 nuclear magnetic resonance studies of the effect of oxygen upon glycolysis in yeast. *Biochemistry* **20**(20): 5871–5880.
22. Thevelein JM and Hohmann S (1995) Trehalose synthase: guard to the gate of glycolysis in yeast? *Trends Biochem. Sci.* **20**: 3–10.

11

Trehalose Energetics in Yeast Spores

Robert G. Shulman and Jan den Hollander

Yale University School of Medicine, Department of Molecular Biophysics and Biochemistry, PO Box 208024, New Haven, CT 06520-8024, USA

The disaccharide α, α-trehalose content of growing vegetative yeast cells is relatively low (1). Only under challenging environments such as dehydration, abrupt exposure to heat or nutrient starvation does it become highly concentrated. In contrast, trehalose is synthesized and stored at high concentrations in yeast spores (2). The formation of trehalose during sporulation as well as its degradation during germination was intensively studied by a series (3–5) of *in vivo* NMR experiments in the 1980s, extended by standard microbiological methods. In these studies the experimental techniques for measuring *in vivo* ^{13}C NMR label flows were developed that have subsequently been applied to study fluxes of intermediary metabolism in rat models and humans, and in micro-organisms. Beyond these methodological developments the intrinsic study of metabolic fluxes and the specific results obtained in those studies have a renewed value today because the interest in trehalose metabolism has been stimulated by the role proposed for trehalose as a protectant in the response of yeast to heat shock (6–8) and particularly in response to drying where it has been shown to maintain membranes (9) and proteins (10).

Evidence for the connection between trehalose synthesis and degradation and the heat shock response has been accumulating rapidly. Upon exposure to heat shock conditions yeast cells increase their trehalose concentration several fold, facilitated by an increase of enzymes in the biosynthetic pathway as mediated

Metabolomics by In Vivo NMR. Edited by R. G. Shulman and D. L. Rothman
 ISBN: 0-470-84719-0

by gene expression (11–13). Accompanying this increase in synthesis there is an increase in activity of the enzyme trehalase which hydrolyzes the disaccharide to glucose (14). Any deficit in the increase as trehalose or in its rate of degradation during recovery results in loss of stress protection. Based upon reports that trehalose directly protects proteins as well as membranes from degradation *in vitro* (15) the *in vivo* increase in trehalose has been interpreted to suggest a similar protection during anhydrobiosis heat shock and starvation *in vivo*. Suggestions have been made as to how both the increase and decrease of trehalose could be consistent with a model in which trehalose at high temperature protects against denaturation whereas its subsequent reduction during return to normal temperature (16) is required in order to allow refolding to the native state by chaperone proteins.

The explanatory basis of the ^{13}C NMR experiments on trehalose metabolism in yeast spores was quite different from the present interest in its role as a protectant. Those studies were directed towards a thermodynamic explanation. Life consumes energy and, although sporulation was a low-energy state, the question addressed was did yeast spores consume energy, and if so, how was energy consumption controlled and what were the fuels? The answer from the NMR studies was that dormant spores did consume energy provided by the catabolism of trehalose (4). The basis of this conclusion was a series of papers on yeast spores that delineated the central role of trehalose in their energetic pathways (3–5, 17). Since these results have not been contradicted in these intervening years, they remain relevant for spores today, and this review may be of wide interest because of the central role being found for trehalose in many processes.

11.1. NMR OF YEAST SPORES

Trehalose accumulates in spores of fungi in high concentrations (18) so that even the natural abundance ^{13}C NMR signal can be observed clearly. During yeast sporulation with acetate as carbon source in a nitrogen starvation medium, large quantities of trehalose are synthesized (up to 35 % of the dry weight). The concentration of this endogenous carbon source decreases slowly during months of dormancy and rapidly upon germination in the presence of glucose. Yeast spores require a carbon source for germination, and indeed when supplied with an exogenous carbon source, such as glucose, the endogenous trehalose is rapidly depleted (5). Trehalose accumulation also occurs in yeast during periods of starvation, and, upon transfer to an environment facilitating growth, the high levels of trehalose are rapidly degraded (1, 19, 20).

At the commencement of the NMR experiments, two possibilities existed for the energetic function of the high concentrations of trehalose in spores. The first, reflecting its sudden decrease during germination, was that the carbohydrate served as a rapid endogenous energy supply for the initial stages of germination, or more generally, to begin the growth cycle (21, 22). A second possibility was that trehalose may be important during the dormant period. The ability of trehalose to provide a source of adenosine triphosphate (ATP) during dormancy might, it was suggested, contribute to maintenance of spore viability. A decrease in trehalose levels in spores of *Neurospora* sp. upon aging was also reported (22). Furthermore in contrast to many expectations that spores were dry, so that chemical reactions could be neglected, the NMR spectra showed (3, 4) narrow lines of several metabolites, indicating the rapid molecular motion characteristic of a liquid medium. From these results it was suggested that the trehalose might serve as an energy supply during dormancy (4). We examined by ^{13}C and ^{31}P NMR the metabolism of trehalose in asci of the yeast strain Y55, which at that time was identified as *Pichia pastoris*, but which now is established to be *Saccharomyces cerevisae*. With this identification in hand the relation of these early NMR results to recent results, primarily on *S. Cerevisae*, will be more apparent.* The rapid, quantitative and non-invasive NMR measurements of trehalose catabolism provided insight into its role as an energy source in yeast spores.

*Dr Cletus P. Kurtzman (Microbial Genomics and Bioprocessing Research Unit, USDA, Peoria, Illinois) kindly points out that there has been quite a bit of confusion about the identity of Y55, including whether the Y55 was really NRRL Y-55. Added to that was the confusion of its original identification, which was *Zygosaccharomyces pastori*. Relatively recently, domains 1 and 2 of 26S

11.2. RESULTS

Utilizing [1-^{13}C] acetate in the sporulation medium, the ^{13}C NMR spectra monitored the synthesis of trehalose. Upon release from dormancy, the ^{13}C NMR signals from the labeled trehalose were measured throughout the germination period with a time resolution of 3 min. Labeled endogenous trehalose and exogenous glucose were both followed in the spores simultaneously over the course of germination. Figure 11.1 shows the ^{13}C NMR spectra of asci containing (^{13}C) trehalose before and after the addition of (1-^{13}C) glucose. The simplicity of the labeling pattern introduced by the acetate label during sporulation is apparent in Figure 11.1(a). Here, in the dormant state before germination, only the C-3 and C-4 peaks of trehalose were labeled, which is expected by incorporation of [1-^{13}C] acetate through the glyoxylate cycle.

Figure 11.1 shows how ^{13}C NMR followed separately the consumption of glucose and trehalose during germination. At time zero (1-^{13}C)glucose was added to a concentration of 74 mM and ^{13}C NMR spectra

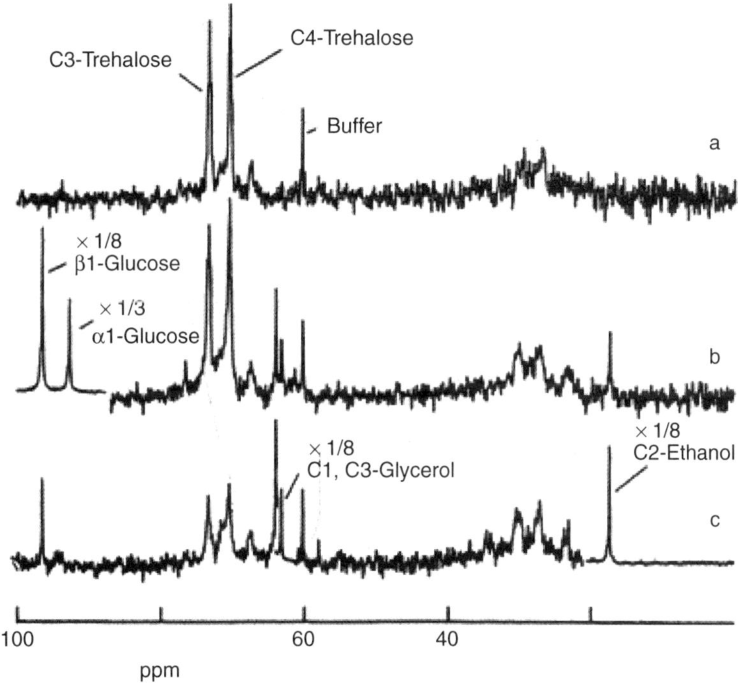

Figure 11.1. ^{13}C NMR spectra of a 20 % suspension of labeled asci before and during germination with 75 mM [1-^{13}C]glucose in 50 mM Tris–5 mM potassium phosphate (pH 7.4) at 25 °C: (a) before germination; (b) accumulations between 4 and 7 min after [1-^{13}C]glucose addition; (c) accumulations between 58 and 61 min (pulse interval, 0.5 s; line broadening, 5 Hz). (Reproduced from Barton JK, den Hollander JA, Hopfield JJ and Shulman RG, *J. Bacteriol.* 1982; **151**(1): 177–185 by permission of the American Society for Microbiology.)

rDNA have been sequenced for many of the strains in their collection, and it was found that for NRRL Y-55 the sequence for this region matches that of *Zygosaccharomyces rouxii*. (CP Kurtzman, personal communication) Dijksterhuis *et al.* (8) further obfuscate the designation by identifying Y55 as *S. cerevisiae*. Professor Johan Thevelein reports (private communication), 'When I came back from the USA I had the identity of Y55 investigated by the CBS in Holland. They stated that it was *Saccharomyces cerevisiae*. The strain was homothallic (switches mating types, like all strains in nature)'. R Young in England used the Y55 strain for studies on meiosis because of its excellent sporulation. He made heterothallic strains by deletion of the HO gene. We have received some of these strains. He concluded that Y55 is *S. cerevisiae* and that the confusion probably arose because Y55 is not the NRRL Y-55 strain.

were accumulated for periods of three minutes. Figure 11.1(b) shows the spectrum accumulated between 4 and 7 min after glucose addition. The (1-^{13}C)glucose peaks from the α and β anomers were very intense and have been reduced by a factor of 8 so as to fit on the scale. In Figure 11.1(b) one can see the appearance of glycerol and ethanol peaks, the glycolytic end products. Figure 11.1(c) shows the spectrum taken between 58 and 61 min after glucose addition, when most of the glucose had been consumed and the trehalose reduced appreciably. In this spectrum, peaks from glycerol and ethanol have been reduced 8-fold. We emphasize that these NMR spectra clearly distinguish between the levels of glucose, as shown by its 1-^{13}C label, and of trehalose, as shown by its 3-^{13}C and 4-^{13}C peaks.

In this experiment, from the peak amplitudes, the ratio of trehalose to glucose catabolized was 0.09. Trehalose, the endogenous carbohydrate, although contributing substantially to the energy upon germination with glucose, still did not serve as the primary energy supply for the germination of dormant spores under the conditions shown in Figure 11.2. Under different conditions (4) the fraction of energy contributed by trehalose can be as high as 30 %. Hence stored trehalose contributes to the energy needed for germination but is not the exclusive fuel, glucose being important in NMR studies.

The relationship between trehalose mobilization and exogenous glucose level can be seen in Figure 11.2. The time course at 25 °C of glucose and trehalose consumption of a low-density ascus suspension is plotted in Figure 11.2(a), and that of ethanol and glycerol production is plotted in Figure 11.2(b). Here

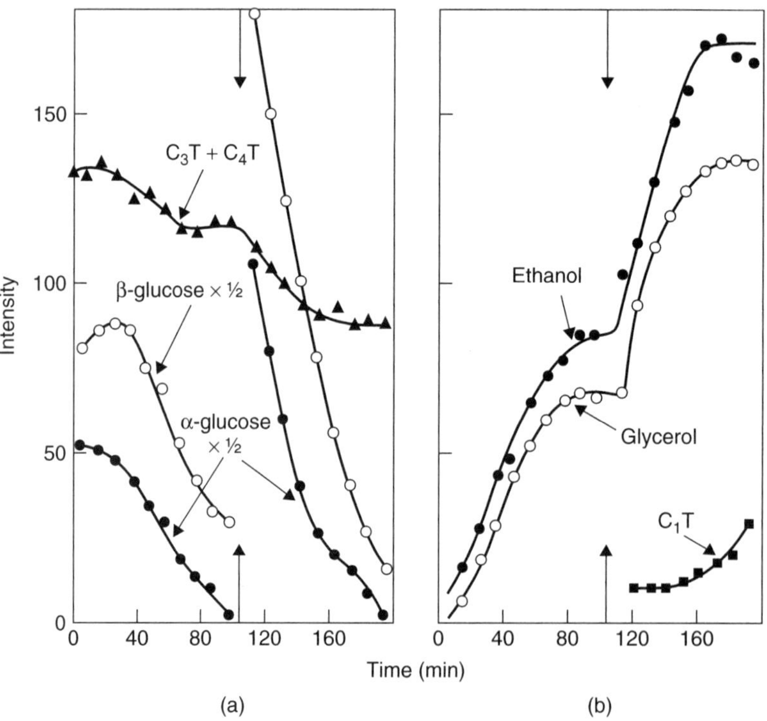

Figure 11.2. Time course at 25 °C during germination at a 4 % suspension of ^{13}C-labeled asci in the absence of nitrogen sources showing (a) glucose and trehalose consumption and (b) the production of ethanol, glycerol and trehalose, based on ^{13}C NMR spectra. ^{13}C NMR spectra were accumulated for 10 min; pulse intervals, 2.0 s. Initial [1-^{13}C]glucose feeding was at a concentration of 4.0 mM. Arrows indicate refeeding with 10 mM [1-^{13}C]glucose. (Reproduced from Barton JK, den Hollander JA, Hopfield JJ and Shulman RG, *J. Bacteriol.* 1982; 151(1): 177–185 by permission of the American Society for Microbiology.)

(1-^{13}C) glucose was added at a low concentration of 4.0 mM. As the glucose was consumed, the trehalose peak intensities decreased by only 12 % and thereafter leveled off as the glucose concentration reached 2.4 mM. After 105 min, the asci solution was replenished by 1-^{13}C glucose to a concentration of 10 mM. Again trehalose consumption started, but when the glucose concentration decreased to ~2.4 mM trehalose consumption halted as previously. It appears that trehalose is only consumed when there is a sufficiently high level of glucose present. The NMR experiments not only determined both trehalose and glucose consumption, but also monitored newly synthesized trehalose between 160 and 200 min after the initial glucose feeding when the glucose was approaching exhaustion [Figure 11.2(b)]. The 1-^{13}C trehalose was obviously synthesized from 1-^{13}C glucose.

Figure 11.2(b) shows that, in the absence of a nitrogen source during early germination, if the glucose level is lowered, thereby halting trehalose mobilization, trehalose may again be stored. Hence low glucose levels have not caused trehalose to be hydrolyzed (as high levels do) but in fact have resulted in new synthesis of trehalose. In the absence of nitrogen sources trehalase activity decreases after the initial activation (23). With nitrogen sources trehalase activity remains high during germination, and no resynthesis of trehalose occurs. Since these experiments were done without nitrogen sources, trehalose resynthesis may be related to the absence of nitrogen sources, rather than to low glucose.

11.3. ATP FROM TREHALOSE PHOSPHORYLATES 2-DEOXYGLUCOSE

Since our experiments demonstrated that the endogenous trehalose was not a primary carbohydrate source for germination, which had been initiated by plentiful glucose, we tested whether trehalose can be mobilized so as to supply ATP in dormant spores. Using ^{13}C NMR and classical extraction methods, we observed that trehalose levels decreased in spores when challenged with 2-deoxyglucose (2DG). A minimum decrease of 19 % in trehalose levels was observed by chemical analyses after incubation with up to 0.5 M 2-deoxyglucose for 18 h (4). Parallel ^{31}P NMR experiments indicate the formation of 2-deoxyglucose 6-phosphate during this period. By quantitating trehalose loss with ^{13}C NMR and 2DG incorporation by ^{31}P NMR in several trials, we found that 3.8 ± 0.4 molecules of 2-deoxyglucose-6-phosphate (2DG-6-P) were phosphorylated for each trehalose molecule consumed (Table 11.1). Since each trehalose molecule consumed through glycolysis should create four equivalents of ATP, which via hexokinase would synthesize four equivalents of 2DG-6-P, this result corresponds to the expected stoichiometric ratio of trehalose catabolism to 2DG-6-P incorporation. Hence these experiments showed first that trehalose is not the dominant energy supply for spore germination in the presence of glucose in a rich medium, and second that trehalose can act as a glycolytic source for ATP in dormant yeast spores

In the first ^{31}P NMR studies of spore suspensions the ATP levels in dormant asci were too low to be detected by ^{31}P NMR, i.e. significantly below 1 mM, in contrast to the normal 5 mM concentration in

Table 11.1. Summary of ascus incubations with 2-DG

Trial	Trehalose consumed (mmol)	2-DG incorporated (mmol)	Moles of 2-DG incorporated/mole of trehalose consumed
1	1.06	4.0	3.8
2	1.02	4.6	4.5
3	1.08	4.3	4.0
4	.9	2.6	2.9
Average			3.8 ± 0.4

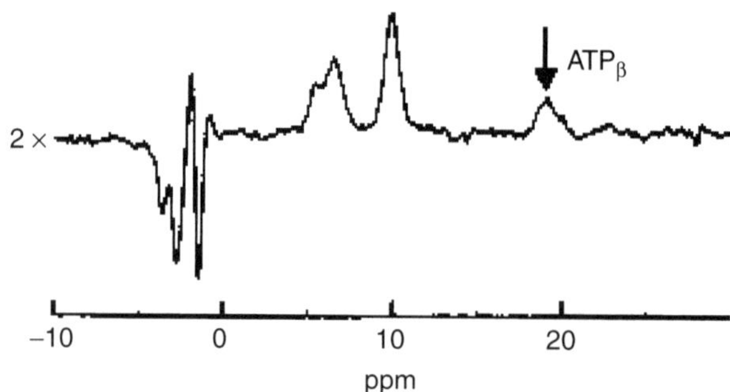

Figure 11.3. ^{31}P NMD difference spectrum of dormant yeast ascospores before and after 2-deoxyglucose feeding. Difference spectrum showed a loss of ATP_{β}, the β phosphate group of ATP corresponding to 0.7 mM. (Reproduced from Thevelein JM, den Hollander JA and Shulman RG, *Proc. Natl Acad. Sci. USA* 1982; **79**: 3503–3507 by permission of the National Academy of Sciences.)

energized yeast cells (24). The ATP content was depleted by feeding an excess of 2DG to the suspension of spores. *In vivo* ^{31}P NMR spectra were taken of dormant spores in the ATP-depleted condition. The low levels of ATP did not allow its concentration to be determined solely by measuring the intensity of the ß-ATP peak, since other ^{31}P resonances, such as from polyphosphates, might contribute. Hence the ATP concentration was determined by differencing the ^{31}P intensity of this peak after all the ATP had been consumed by excess 2DG. From the difference spectrum (Figure 11.3), the dormant spores were calculated to have an intracellular ATP concentration of about 0.7 mM.

To determine whether the enhanced trehalose breakdown was correlated with increased activity of trehalase, kinetic measurements were made of trehalase activity in spore extracts (5). Within the first minutes after addition of glucose and phosphate, a sudden 10-fold increase of the trehalase activity occurred. Maximal activity was seen after about 10 min; subsequently the activity gradually declined back to low values. As was found for trehalose breakdown during germination, glucose, nitrogen and phosphate were necessary for maximal activation of trehalase. Glucose alone caused only a small activation. However, addition of 1 mM phosphate was enough to obtain nearly maximal activation, which was reached with ~5 mM phosphate and ~100 mM glucose (5).

In addition to this increase in activity with germination, there were differences in the control properties found between the enzymes from dormant and germinating spores. Trehalase from dormant spores was very sensitive to inhibition by ATP: it was strongly inhibited by ATP at a concentration of ~0.5 mM (Figure 11.4). On the other hand, trehalase from germinating spores was not inhibited by ATP up to much higher ATP concentrations (there are two forms of trehalase – an acid trehalase in the vacuoles, which is always active, and a cytosolic form, which is activated by the phosphorylation cascade). Therefore, in the dormant spore we may have been looking mainly at the acid trehalase, whereas in the germinating spore presumably the cytosolic enzyme kicked in. A discussion is found in Kopp *et al.* (25).

The level of trehalase activity immediately after the induction of germination was larger than expected from the extremely slow mobilization of trehalose in dormant spores. The *in vitro* experiments on the regulatory properties of trehalase revealed an ATP inhibition of the enzyme in dormant spores (Figure 11.4). The ATP concentration of ~0.7 mM measured in dormant spores by ^{31}P NMR spectroscopy is close to the K_I value measured in the spore form of trehalase, and would strongly inhibit trehalase activity (Figure 11.4). Therefore, it is proposed that the slow mobilization of trehalose in dormant spores is caused both by the inherently low activity of trehalase and by stringent ATP feedback inhibition. The ATP control shows that increased consumption of ATP in the dormant spore should cause a concomitant stimulation of trehalose

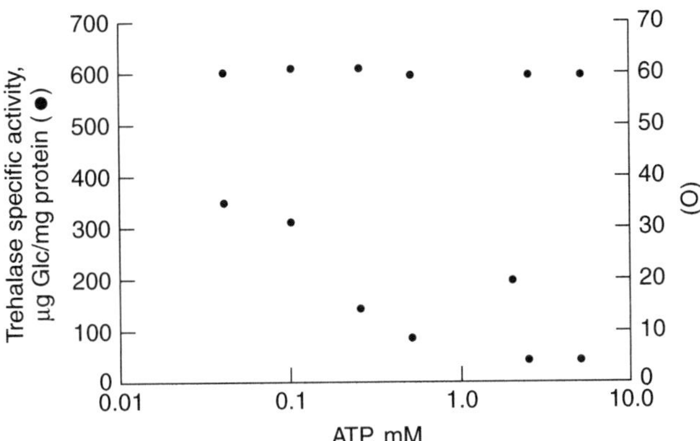

Figure 11.4. Trehalase activity in dormant (○) and germinating (•) spores showing how ATP controls trehalase activity during dormancy but not during germination. Germinating spores, 15 min incubation in complex medium (1 % yeast extract, 2 % bactopeptone, and 100 mM glucose). (Reproduced from Thevelein JM, den Hollander JA and Shulman RG, *Proc. Natl Acad. Sci. USA* 1982; **79**: 3503–3507 by permission of the National Academy of Sciences.)

breakdown. The existence of such a correlation was shown above when 2DG was converted into 2DG-6-P with concomitant decrease in trehalose content. Apparently, trehalose was used to maintain the ATP constant at ~0.5 mM. The stoichiometry of the reactions involved indicated that the amount of ATP that could be produced from the amount of trehalose broken down equaled the amount of ATP needed to phosphorylate deoxyglucose. It can be concluded from these results that the trehalose reserve in dormant spores can be used to supply energy and that its rate of glycolysis is regulated by the demand for ATP, with the regulated enzyme trehalase serving to maintain a low (~0.5 mM), constant level of ATP.

In summary in the dormant spores, low trehalase activity and strong ATP inhibition are responsible for maintaining the slow mobilization of trehalose. The slow mobilization might be adjusted to the metabolic needs through the effects of intracellular pH and the cAMP level upon trehalase activation (26, 27). However, regardless of the mechanism establishing trehalase activity in spores, the enzyme acts in a pathway providing energy. During early germination, the switch toward high trehalase activity and the loss of ATP inhibition are responsible for the rapid breakdown of trehalose.

11.4. DISCUSSION

This study revealed much about the energetic pathways of trehalose metabolism and the novel nature of their control. Unfortunately these results were overlooked by a paper in 1990 which denied the energetic function of trehalose (28). Since the paper entitled 'Trehalose in yeast, stress protectorant rather than reserve carbohydrate' misunderstood our earlier studies, claiming they came to a conclusion opposite to our actual findings (4), Wiemken's conclusions are re-examined here.

As reviewed above, Barton *et al.* (4) considered two possible usages for trehalose energetics. It had previously been argued that, since trehalose was hydrolyzed upon germination, perhaps its energy was needed at that time to support the many accompanying cellular changes. Yet the ^{13}C NMR measurements of the relative energy from trehalose and exogenous glucose showed that when trehalose was almost completely metabolized it had only contributed ~9 % to the total energy consumption during germination. This led Barton *et al.* to the alternative conclusion that trehalose energy was needed during dormancy, as reviewed above. Wiemken (28), quoting only part of the Barton paper and misreading its conclusion, wrote that 'In gross contradiction to the alleged reserve function of trehalose... ^{13}C NMR revealed that the trehalose degraded by germinating spores of the yeast *Pichia pastoris* contributed but 9 % of the glucose consumed during the early germination phase'. In contradiction to Barton *et al.*'s data and conclusion summarized above, Wiemken drew the opposite inferences from their results, claiming that, since trehalose is not used for energy during germination, it is simply not used for energy – period. He then concluded 'We strongly plead for dismissal of the proposition that trehalose in yeast (and other fungi) has a primary role as reserve carbohydrate.' He goes on to say that it protects against 'adverse conditions' and, despite the conclusion drawn from narrow NMR lines (3) of a mobile spore interior, he suggested that trehalose protects dormancy by slowing down metabolism.

The protection of macromolecular structure by trehalose has been based upon *in vitro* physical chemistry experiments (29, 30) with which we have no quarrel. Many *in vivo* experiments are being interpreted in these terms which support this role for trehalose during stress. Of course there is no reason why trehalose may not serve more than one function during stress. However we feel that its energetic role has been well established by Barton's experiments and unnecessarily neglected for reasons given above. Other arguments strongly supporting an energetic role for trehalose exist, such as the presence of only trehalose as a reserve carbohydrate in certain fungal spores capable of germinating in distilled water without any exogenous carbon source. It has also been overlooked that the fluctuations in the trehalose content in yeast in many instances run parallel to those of glycogen, a well-established reserve carbohydrate without stress-protectant function.

In vegetative yeast, as studied by ^{31}P NMR, the ATP levels are ~5 mM and remain constant over a wide range of conditions. Similarly mammalian cells in liver, muscle, brain and kidney maintain constant ATP levels of ~5 mM. Steady states which maintain submillimolar concentrations of ATP are very rare. In Tetrahymena at temperatures where heat shock proteins are synthesized, ATP concentrations significantly below 5 mM are maintained (31). Similar reductions in ATP levels, with concomitant increase in AMP, occur in heat shock of hepatocytes (32). Mammalian cells maintain constant ATP by the buffering action of phosphorous creatine via the rapid creatine kinase enzyme. Therefore the low ATP concentration in dormant spores is an unusual state of life: between the usual several millimolar and the state where ATP is zero and the cell is dead. The control of this low level in spores is elucidated by the experiments described above.

The glycolytic pathway, whose first step is the allosteric enzyme, trehalase, with an activity controlled by the product, ATP, is a classic case of the demand control of flux (33). Increasing the demand for ATP with endogenous 2-DG caused an increase in the trehalose catabolism. As Hofmeyr and Cornish-Bowden (33) showed, if the flux through a pathway is controlled by the demand for the product, then an allosteric enzyme at an early stage, as trehalase in yeast spores has been shown to be, responds to accommodate the flux but is not controlling it. Rather it is increasing its activity so as to maintain ATP and other metabolites in the pathway constant. It is regulating the intermediates so as to maintain homeostasis – keeping their concentrations constant in the face of changing fluxes. The inference we draw is that the normal complement of enzymes that consume ATP in the dormant spore creates a demand for ATP, which is satisfied by trehalase hydrolysis of trehalose, thereby maintaining approximately 0.5 mM ATP.

The consequences of an order of magnitude lower ATP concentrations when compared with vegetative yeast can be envisaged qualitatively even without additional information about the enzymes involved. Some, perhaps all, of the ATP consuming enzymes in vegetative yeast, accustomed to ~5 mM ATP, are operating in the vicinity of K_m. Decreasing the ATP concentration will decrease their rates. Hence this single metabolic change can account for some of the reduced metabolism characteristic of spores – perhaps it is responsible for a large fraction of that decrease. This does not mean that other mechanisms, e.g. altered gene expression or compartmentation of metabolites and enzymes, do not contribute to the modified metabolism, but it does suggest a strong contribution of ATP concentration.

Given the several analogies between trehalose metabolism in yeast spores and cells, we suggest that its cellular energetic functions in the cells might warrant study, particularly since it is known that heat shock in tetrahymena is accompanied by a lowered energy state where ATP levels are decreased (31). We do not claim that trehalose protects yeast during heat shock because of its ability to provide energy. Not even if experiments were to show that it is catabolized during heat shock would that disallow its proposed role as a protectant of macromolecular structure. However we do claim that the energetic role of trehalose cannot be disregarded on the basis of the NMR experiments in yeast spores, which have shown that trehalose acts as an energy source.

Acknowledgments

We thank Pieter van Eijsden, Clifford Slayman and Johan Thevelein for helpful suggestions.

REFERENCES

1. Van der Plaat JB (1974) Cyclic 3′,5′-adenosine monophosphate stimulates trehalose degradation in baker's yeast. *Biochem. Biophys. Res. Commun.* **56**: 580–587.
2. Kane SM and Roth RM (1974) Carbohydrate metabolism during ascospore development in yeast. *J. Bacteriol.* **118**: 8–14.
3. Barton JK, den Hollander JA, Lee TM, MacLaughlin A and Shulman RG (1980) Measurement of the internal pH of yeast spores by ^{31}P nuclear magnetic resonance. *Proc. Natl Acad. Sci. USA* **77**(5): 2470–2473.
4. Barton JK, den Hollander JA, Hopfield JJ and Shulman RG (1982) ^{13}C nuclear magnetic resonance study of trehalose mobilization in yeast spores. *J. Bacteriol.* **151**(1): 177–185.
5. Thevelein JM, den Hollander JA and Shulman RG (1982) Changes in the activity and properties of trehalase during early germination of yeast ascopores: correlation with trehalose breakdown as studied by *in vivo* ^{13}C NMR. *Proc. Natl Acad. Sci. USA* **79**: 3503–3507.
6. Panek AD and Panek AC (1990) Metabolism and thermotolerance function of trehalose in Saccharomyces: a current perspective. *J. Biotechnol.* **14**: 229–238.
7. Elbein AD, Pan YT, Pastuszak I and Carrol D (2003) New insights on trehalose: a multifunctional molecule. *Glycobiology* **13**(4): 17R–27R.
8. Dijksterhuis J, van Driel KG, Sanders MG, Molenaar D, Houbraken JAMP, Samson RA and Kets EPW (2002) Trehalose degradation and glucose efflux precede cell ejection during germination of heat-resistant ascospores of Talaromyces macrosporus. *Arch. Microbiol.* **178**: 1–7.
9. Crowe JH, Crowe LM, Oliver AE, Tsvetkova N, Wolkers W and Tablin F (2001) The trehalose myth revisited: introduction to a symposium on stabilization of cells in the dry state. *Cryobiology* **43**: 89–105.
10. Carpenter JF, Martin B, Crowe LM and Crowe JH (1987) Stabilization of phosphofructokinase during air drying with sugars and sugar/transition metal complexes. *Cryobiology* **24**: 455–464.
11. Voit EO (2003) Biochemical and genomic regulation of the trehalose cycle in yeast: review of observations and canonical model analysis. *J. Theor. Biol.* **223**: 55–78.
12. Nwaka S and Holzer H (1998) Molecular biology of trehalose and the trehalases in the yeast *Saccharomyces cerevisiae*. *Prog. Nucl. Acid Res. Mol. Biol.* **58**: 197–237.
13. Winderickx J, de Winde JH, Crauwels M, Hino A, Hohmann S, Van Dijck P and Thevelein JM (1996) Regulation of genes encoding subunits of the trehalose synthase complex in *Saccharomyces cerevisiae*: novel variations of STRE-mediated transcription control? *Mol. Gen. Genet.* **252**: 470–482.
14. Wera S, DeSchrijver E, Geyskens I, Nwaka S and Thevelein JM (1999) Opposite roles of trehalase activity in heat-shock recovery and heat-shock survival in *Saccharomyces cerevisiae*. *Biochem. J.* **343**: 621–626.
15. Crowe JH, Hoekstra FA and Crowe LM (1992) Anhydrobiosis. *A. Rev. Physiol.* **54**: 579–599.
16. Singer MA and Lindquist S (1998) Thermotolerance in Saccharomyces cerevisiae: the Yin and Yang of trehalose. *Trends Biotechnol.* **16**: 460–468.

17. Thevelein JM, den Hollander JA and Shulman RG (1984) Trehalase and the control of dormancy and induction of germination in fungal spores. *Trends Biochem. Sci.* **9**(11): 495–497.
18. Thevelein JM (1984) Regulation of trehalose mobilization in fungi. *Microbiol Rev* **48**: 42–59.
19. Lillie SH and Pringle JR (1980) Reserve carbohydrate metabolism in *Saccharomyces cerevisiae*. Response to nutrient limitation. *J. Bacteriol.* **143**: 1384–1394.
20. Hirimburegama K, Durnez P, Keleman J, Oris E, Vergauwen R, Mergelsberg H and Thevelein JM (1992) Nutrient-induced activation of trehalase in nutrient-starved cells of the yeast *Saccharomyces cerevisiae*. cAMP is not involved as second messenger. *J. Gen. Microbiol.* **138**: 2035–2043.
21. Panek AD (1963) Function of trehalose in baker's yeast (*Saccharomyces cerevisiae*). *Arch. Biochem. Biophys.* **100**: 422–425.
22. Lingappa BT and Sussman AS (1959) Endogenous substrates of dormant, activated and germinating ascospores of *Neurospora tetrasperma*. *Plant Physiol.* **34**: 466–472.
23. Thevelein JM and Jones KA. (1983) Reversibility characteristics of glucose-induced trehalase activation associated with the breaking of dormancy in yeast ascospores. *Eur. J. Biochem.* **15:136**(3): 583–587.
24. Gillies RJ, Ugurbil K, den Hollander JA and Shulman RG (1981) ^{31}P NMR studies of intracellular pH and phosphate metabolism during cell division cycle of *Saccharomyces cerevisiae*. *Proc. Natl Acad. Sci. USA* **78**(4): 2125–2129.
25. Kopp M, Muller H and Holzer H (1993) Molecular analysis of the neutral trehalase gene from *Saccharomyces cerevisiae*. *J. Biol. Chem.* **268**(7): 4766–4774.
26. Thevelein JM, Beullens M, Honshoven F, Hoebeeck G, Detremerie K, den Hollander JA and Jans AWH (1987) Regulation of the cAMP level in the yeast *Saccharomyces cerevisiae*: intracellular pH and the effect of membrane depolarizing compounds. *J. Gen. Microbiol.* **133**: 2191–2196.
27. Thevelein JM and de Winde JH (1999) Novel sensing mechanisms and targets for the cAMP – protein kinase A pathway in the yeast *Saccharomyces cerevisiae*. *Mol. Microbiol.* **33**: 904–918.
28. Wiemken A (1990) Trehalose in yeast, stress protectant rather than reserve carbohydrate. *Antoine Van Leeuwenhoek* **58**(3): 209–217.
29. Iwahashi H, Obuchi K, Fujii S and Komatsu Y (1995) The correlative evidence suggesting that trehalose stabilizes membrane structure in the yeast Saccharomyces cerevisiae. *Cell. Mol. Biol. (Noisy-le-grand)*, **41**(6): 763–769.
30. Thevelein JM, 1996, Regulation of trehalose metabolism and its relevance to cell growth and function. In: *The Mycota: a Treatise on the Biology of Fungi with Emphasis on Systems for Fundamental and Applied Research*, Esser K and Lemke PA (eds), Vol. III, Biochemistry and Molecular Biology, Chap. 19, pp. 395–420. Springer: Berlin.
31. Findly RC, Gillies RJ and Shulman RG (1983) *In vivo* ^{31}P NMR reveals lowered ATP during heat shock of *tetrahymena*. *Science* **219**: 1223–1225.
32. Corton JM, Gillespie JG and Hardie DG (1994) Role of AMP-activated protein kinase in the cellular stress response. *Curr. Biol.* **4**: 315–324.
33. Hofmeyr, J and Cornish-Bowden, A. (1990) Quantitative assessment of regulation in metabolic systems. *FEBS Lett.* **476**: 47–51.

12

Metabolic Networks in the Liver by ^2H and ^{13}C NMR

A. Dean Sherry

Professor of Chemistry, University of Texas at Dallas, Richardson, TX 75083 and Professor of Radiology, University of Texas Southwestern Medical Center, Dallas, TX 75235, USA

Craig R. Malloy

Professor of Radiology and Internal Medicine, University of Texas Southwestern Medical Center, Dallas, TX 75235, and VA North Texas Health Care System, USA

Metabolomics by In Vivo NMR. Edited by R. G. Shulman and D. L. Rothman
 ISBN: 0-470-84719-0

12.1. INTRODUCTION

The liver plays the central role in whole body fuel metabolism through multiple interacting pathways. It has the dual capacity of releasing glucose from glycogen or synthesizing glucose from glycerol or from a wide range of precursors that pass through the tricarboxylic acid (TCA) cycle. During postprandial hyperglycemia, gluconeogenesis is suppressed to some extent, glycogen is stored and wide fluctuations in plasma glucose concentrations are avoided. The liver oxidizes fatty acids for energy. Yet, during periods of nutritional abundance, it also converts excess carbohydrate and protein into fatty acids. Thus, a fundamental feature of liver metabolism is its capacity to rapidly shift its metabolic activity. For this reason, it is essential to describe the metabolic profile of the liver in terms of fluxes through relevant metabolic pathways including glycogenolysis, gluconeogenesis and fuel oxidation in the TCA cycle. Flux measurements through multiple interacting pathways are difficult in isolated hepatocytes and even more challenging *in vivo* because of the intricate connections between catabolic and anabolic pathways. Although radiotracers offer superb sensitivity, the experimental information required to measure fluxes in complex metabolic networks exceeds the information yield from a single radiotracer experiment. Thus, even though the fundamentals of metabolism as we currently know them were originally derived using of radiotracers, alternative methods such as NMR will be needed to fully understand metabolism in intact tissues.

NMR spectroscopy in combination with stable isotope tracer molecules offers the potential of providing new insights into control of metabolism in humans. ^{13}C NMR in particular is attractive because it is relatively easy to deliver highly enriched tracer molecules containing either single or multiple ^{13}C enriched carbons. ^{13}C NMR can be used to measure absolute TCA cycle flux, anaplerosis, gluconeogenesis, O_2 consumption and substrate oxidation in human liver. One approach is to use whole-body, high-field magnetic resonance spectroscopy (MRS) techniques to follow metabolic processes directly as they happen *in vivo* (Rothman *et al.*, 1991; Jucker *et al.*, 1998). These tools are powerful but unfortunately high-field MRS systems are not available at most medical centers. A complementary approach is to use analytical NMR spectroscopy in combination with stable isotope tracers to derive metabolic flux data from molecules easily obtained from plasma, urine or tissue biopsies. Such methodology has matured to a point where high-throughput analysis of metabolism in large patient populations is now eminently feasible. The state of the art of this methodology is reviewed in this article.

12.2. ^{13}C TRACERS FOR ANALYSIS OF METABOLIC NETWORKS

For more than 60 years, hepatic metabolism has been studied using ^{14}C tracer molecules. Here, the distribution of label in cycle intermediates is typically fit to mathematical models of varying complexity to measure relative flux through critical pathways such as substrate oxidation and gluconeogenesis. ^{13}C NMR was first used to follow intermediary metabolism in *Saccharomyces cervesiae* (Eakin *et al.*, 1972). The first ^{13}C NMR kinetic analysis of the TCA cycle, reported a decade later, used ^{13}C fractional enrichments in glutamate carbons (analogous to ^{14}C-specific activity measurements) to calculate TCA cycle flux in the isolated heart (Chance *et al.*, 1983). It was noted in that early paper that each glutamate carbon resonance appeared as complex multiplets due to ^{13}C–^{13}C spin–spin coupling and that these multiplets were a consequence of turnover of the TCA cycle. At approximately the same time, it was realized that this complexity reflected the activities of various pathways feeding into the TCA cycle and the development of tools for analysis of these multiplets was initiated (Malloy *et al.*, 1987, 1988). In addition to the extra information provided by these complex multiplets, there are also practical advantages of using ^{13}C as a metabolic tracer. All concerns related to radiation exposure are eliminated so experiments can be performed in virtually any laboratory or clinical environment. Perhaps less obvious, various ^{13}C-enriched compounds are now commercially available, including compounds with complex labeling patterns such as fatty acids having an enriched ^{13}C in alternate carbons. Such complex labeling patterns are difficult if not impossible to obtain for ^{11}C or ^{14}C isotopes. There are also advantages that result from the NMR method itself. For example, purification, chemical degradation or formation of metabolic derivatives is often not required so that all information can often be collected in a single spectrum.

12.3. METHODS: ^{13}C AND ^{2}H NMR

^{13}C has a favorable nuclear spin ($I = 1/2$) like that of the proton, and it displays a much wider chemical shift range (200 vs 10 ppm). This permits resolution of ^{13}C resonances in molecules that are structurally quite similar (such as amino acids) even at low field strengths where the corresponding ^{1}H resonances often overlap. However, the NMR receptivity of ^{13}C is only 1.76×10^{-4} compared with ^{1}H for an equal number of nuclei, so this is the primary reason why ^{13}C studies are performed at high magnetic fields. ^{13}C also has a low natural abundance (~1.1 %), so selective labeling of certain carbons in a tracer molecule of interest makes it relatively straightforward to trace any metabolic pathway of interest even in complex biological systems. ^{13}C nuclei must also be spin-decoupled to remove the complexities of proton–carbon spin–spin coupling. This results in both spectral simplification and improved sensitivity due to the gain in intensity by up to a factor of 3 due to the nuclear Overhauser enhancement (nOe). This combined effect makes ^{13}C reasonably favorable for NMR experiments on biological samples. Although the nOe can approach the maximal theoretical value (3-fold) for aliphatic-type carbons (carbon atoms with directly bonded protons), carboxyl- and carbonyl-type carbons may also experience a small nOe due to interactions with protons two or more bonds away. Depending upon the sample, the magnetic field and pulsing conditions, ^{13}C can approach the sensitivity of ^{31}P for NMR detection. ^{13}C resonance intensities are influenced by both T_1 and nOe. Although ^{13}C–^{13}C dipolar coupling provides a relaxation pathway for adjacent ^{13}C nuclei, this interaction is quite weak, about 100-fold smaller than the ^{1}H–^{13}C dipolar interaction, and consequently its influence on T_1 is ordinarily overwhelmed by the much stronger ^{1}H–^{13}C dipolar coupling, especially in carbons with directly bonded protons. Hence, relative multiplet intensities within any single resonance from a protonated carbon atom should not be influenced by pulsing rate. When comparing different carbon nuclei within the same molecule, and particularly when comparing protonated vs carbonyl carbons, the relative resonance intensities are typically sensitive to both pulsing rate and nOe.

The second stable isotope discussed in this chapter is ^{2}H, a nucleus with a quadrupole moment and a spin of 1. Consequently, the relaxation times of ^{2}H are short and the NMR lines are relatively broad compared with ^{1}H, ^{13}C and other spin 1/2 nuclei. The natural abundance of ^{2}H, 0.015 %, is even lower than that of ^{13}C, so enrichment to even as little as 0.5 % results in a ~30-fold increase in signal above background. Although the chemical shift dispersion of ^{2}H is small (~10 ppm for common metabolites), most ^{2}H spectra derived from tissue samples are relatively uncrowded because not all metabolites become enriched in ^{2}H in a typical biological experiment. The gyromagnetic ratio of ^{2}H is also quite low so its NMR receptivity is only 1.45×10^{-6} compared with the ^{1}H. Given the rapid development of high sensitivity, low-volume HPLC-type probes and perhaps even cryo-probes, ^{2}H NMR is feasible on most standard analytical NMR spectrometers. ^{1}H decoupling is also required but this is not technically difficult for an *in vitro* experiment.

12.4. RELATING ISOTOPOMERS TO METABOLISM: ENTRY OF ^{13}C-ENRICHED ACETYL-COA INTO THE TCA CYCLE

The word 'isotopomer', a contraction of the words isotope and isomer, refers to molecules that differ from one another solely on the basis of isotope distribution (London, 1988). The maximum number of ^{13}C isotopomers in a molecule with n carbons is 2^n. Thus, glutamate with five carbons (symbolized as C1, C2, C3, C4 and C5 from the α-carboxyl end) can form 32 possible isotopomers, usually described by a numerical listing of each site of ^{13}C enrichment. All unnumbered sites are assumed to be at natural abundance levels of ^{13}C (1.1 %). As an example, [1,4,5-^{13}C]glutamate indicates ^{13}C enrichment in carbons 1, 4 and 5 while carbons 2 and 3 are ^{12}C (98.9 %). The relative concentrations of these 32 glutamate isotopomers (or 64 glucose isotopomers) are sensitive to the ^{13}C labeling pattern of compounds entering the liver and to fluxes through competing pathways. Other than very simple molecules, neither ^{13}C NMR nor any other carbon tracer technique directly measures the concentration of every isotopomer present in a sample. Rather, ^{13}C NMR measures the relative concentrations of *groups of isotopomers* within any metabolite pool such as glucose or glutamate. Nevertheless, in most cases even ratios of *groups of isotopomers* provide sufficient information for a determination of relative flux values. The term ^{13}C isotopomer analysis refers to the analytic process of relating *groups of isotopomers* by analysis of metabolite multiplets (described below) that appear in a ^{13}C spectrum to obtain relative fluxes through complex, interrelated biochemical pathways.

It is relatively easy to predict the distribution of ^{13}C in all TCA cycle intermediates if the ^{13}C isotopomer distribution in the 2-carbon acetyl group of acetyl-CoA (coenzyme A) is known. Consider the example of Figure 12.1, where only the methyl carbon of acetyl-CoA is enriched with ^{13}C (derived from glucose, ketones or free fatty acids). Condensation of [2-^{13}C]acetyl-CoA with unenriched OAA, a reaction catalyzed by citrate synthase, produces [4-^{13}C]citrate, which is subsequently converted into [4-^{13}C]α-KG and [4-^{13}C]glutamate on the first turn of the cycle. This would be detected by ^{13}C NMR as a singlet in the glutamate C4 resonance at 34 ppm. [4-^{13}C]α-KG continuing through cycle pathways produces [3-^{13}C]succinyl-CoA and subsequently equal amounts of [2-^{13}C]- and [3-^{13}C]succinate, fumarate, malate and oxaloacetate after randomization of label at the symmetric intermediates, succinate and fumarate. Once [2-^{13}C]oxaloacetate and [3-^{13}C]oxaloacetate are formed, these molecules have a

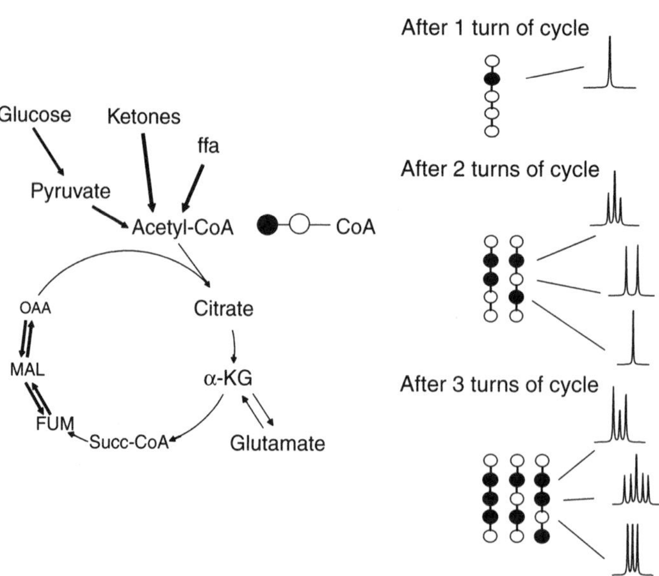

Figure 12.1. Entry of methyl ^{13}C-enriched acetyl-CoA (derived from glucose, ketone bodies or free fatty acids) sequentially enriches all intermediates of the cycle in predictable positions. The right column shows the glutamate ^{13}C isotopomers and the corresponding ^{13}C NMR spectra after turns 1–3. After several turns, depending upon pool size, the isotopomer populations reach steady state and the NMR spectra no longer change.

certain probability of condensing with another [2-^{13}C]acetyl-CoA (proportional to the fractional enrichment of [2-^{13}C]acetyl-CoA) to yield [2,4-^{13}C]citrate, [3,4-^{13}C]citrate which are subsequently converted to [2,4-^{13}C]α-KG, [3,4-^{13}C]α-KG and [2,4-^{13}C]glutamate, [3,4-^{13}C]glutamate. The first of these would appear in the ^{13}C NMR spectrum as singlets in both glutamate C2 and C4 while the second would appear as a doublet (with $J_{34} = 34$ Hz) in both glutamate C3 and C4. Thus, the areas of the singlet vs doublet components in glutamate C4 directly report the relative amounts of the two isotopomers, [2,4-^{13}C]glutamate and [3,4-^{13}C]glutamate. After a third turn through the cycle, the glutamate C2, C3 and C4 resonances would have singlet, doublet and, in the case of C3, triplet components. With further cycle turnover, the glutamate multiplets continue to evolve in intensity and eventually no longer change. This, by definition, is called isotopic steady state because the amount of ^{13}C entering the cycle on each turn is exactly matched by an equal amount of $^{13}CO_2$ leaving the cycle on each turn. In a perfused heart, this takes about 30 min and, in a perfused liver, somewhat longer. At this point, the population of ^{13}C isotopomers in glutamate and all other intermediates is a constant and the information encoded by the multiplets in the ^{13}C NMR spectrum of glutamate reflects metabolism in the TCA cycle.

Now if one considers a more complex but realistic example of metabolism where multiple ^{13}C-enriched substrates are available for oxidation, then the spectra appear slightly more complicated but remain relatively easy to interpret. Consider a situation where unlabeled glucose, [3-^{13}C]lactate and [U-^{13}C]fatty acids are all available for oxidation. In this case, the glucose yields only unenriched acetyl-CoA, [3-^{13}C]lactate yields only [2-^{13}C]acetyl-CoA and [U-^{13}C]fatty acids yield only [1,2-^{13}C]acetyl-CoA, and the proportion of each substrate that contributes to acetyl-CoA and enters the TCA cycle will determine the final population of all possible isotopomers present at isotopic steady-state. The 24 possible glutamate isotopomers that could be formed from this mixture of substrates are shown in Figure 12.2 along with a ^{13}C NMR spectrum of a heart perfused with this substrate mix to isotopic steady state. One can then easily deconvolute each

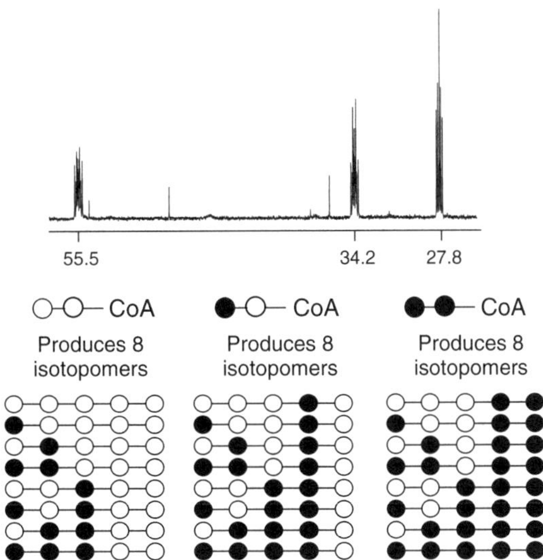

Figure 12.2. ^{13}C NMR spectrum of a heart after perfusion to steady-state with a mixture of unlabeled glucose, [3-^{13}C]lactate and [U-^{13}C]fatty acids. The three major resonances in the spectrum may be assigned to glutamate C2, C4 and C3 (left to right, respectively). All 24 possible ^{13}C isotopomers of glutamate are shown. In this case, unlabeled glucose could contribute eight isotopomers, [3-^{13}C]lactate could contribute eight isotopomers, and [U-^{13}C]fatty acids could contribute another eight isotopomers. A deconvolution of the multiplets that make up each glutamate resonance provides a measure of groups of isotopomers.

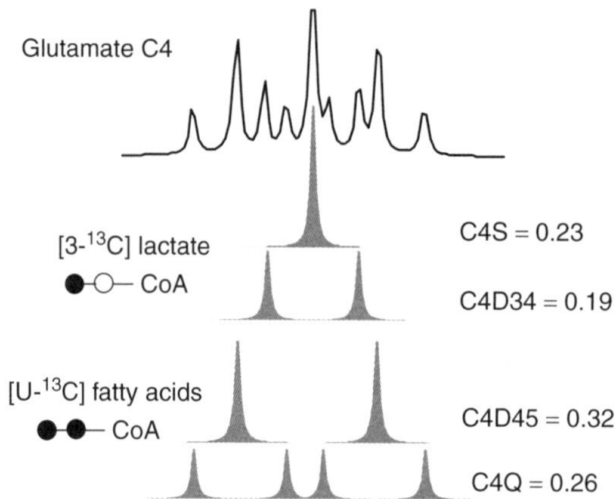

Figure 12.3. Deconvolution of a typical glutamate C4 resonance into a singlet (S), two doublets (D34 and D45), and a doublet-of-doublets or quartet (Q). The area of each multiplet is expressed as a percentage of the total C4 resonance area.

glutamate resonance into its individual multiplet components and use this information to perform a complete isotopomer analysis. An example of deconvolution of a glutamate C4 resonance into individual multiplet components in shown in Figure 12.3. In this example, the central singlet (C4S) makes up 23 % of the total C4 resonance area, the doublet with coupling constant $J_{34} = 34$ Hz (C4D34) contributes 19 %, the doublet with coupling constant $J_{45} = 52$ Hz (C4D45) contributes 32 %, and finally the doublet-of-doublets or quartet (C4Q) contributes another 26 %. As these coupling constants and chemical shifts are indeed constants for glutamate (as long as the pH of the sample is maintained near 7), the ^{13}C spectra look very much alike in every metabolic experiment with only variations in the relative areas of the individual multiplet components. Equations that relate such glutamate multiplet areas to metabolic fluxes (published elsewhere in Malloy *et al.*, 1988, 1990a; Sherry *et al.*, 2004) may be used to derive the contributions of all labeled substrates to acetyl-CoA (the F_{Ci} parameters), the flux of all labeled and unlabeled carbons entering the TCA cycle as an anaplerotic reaction, and flux through all pathways related to the TCA cycle such as pyruvate cycling.

If, however, one is primarily interested in evaluating the contribution of carbohydrate vs fatty acids to energy production, then a much simpler isotopomer analysis can be performed. For example, consider the two glutamate C4 resonances from spectra of heart extracts after perfusion to isotopic steady-state physiological levels of unlabeled glucose, [3-^{13}C]lactate and [U-^{13}C]fatty acids (Figure 12.4). The top spectrum came from the heart of a control rat while the bottom spectrum came from the heart of a spontaneously hypertensive rat (SHR). The peaks that can only arise from oxidation of [3-^{13}C]lactate are labeled with an 'L' and the peaks that can only arise from oxidation of [U-^{13}C]fatty acids are labeled with an 'F'. The contribution of unlabeled glucose to acetyl-CoA is not reflected directly in the glutamate C4 resonance but this contribution can be determined by difference [see Equation (12.3)]. By measuring the areas of these glutamate C4 multiplets and the total areas of the C4 and C3 resonances, one can easily determine the substrate contributions to acetyl-CoA using the following simple equations (Malloy *et al.*, 1990b).

$$\text{C4Q} * (\text{C4/C3}) = F_{C3} \tag{12.1}$$

$$\text{C4D34} * (\text{C4/C3}) = F_{C2} \tag{12.2}$$

$$1 - F_{C2} - F_{C3} = F_{C0} \tag{12.3}$$

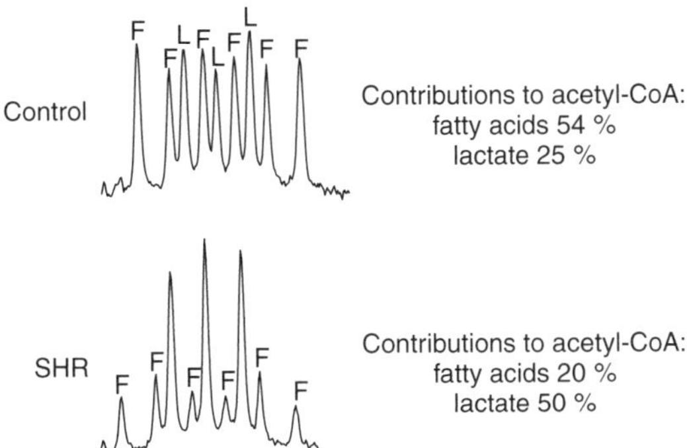

Figure 12.4. Expanded glutamate C4 resonances of hearts from control and spontaneously hypertensive rats. The simple isotopomer analysis described in the text provided the contributions of [U-^{13}C]fatty acids (peaks labeled 'F') and [3-^{13}C]lactate (peaks labeled 'L') to acetyl-CoA.

where C4/C3 is simply the ratio of the total glutamate C4 resonance area divided by the total glutamate C3 resonance area, F_{C2} is defined as the fraction of acetyl-CoA enriched only in the methyl carbon (in this case, derived from [3-^{13}C]lactate), F_{C3} is defined as fraction of acetyl-CoA enriched in both the carbonyl and the methyl carbons (in this case derived from [U-^{13}C]fatty acids), and F_{C0} is the fraction of acetyl-CoA with natural abundance ^{13}C in both the carbonyl and methyl carbons (in this case coming from either exogenous unlabeled glucose or endogenous substrates).

An analysis of the spectra shown in Figure 12.4 using Equations (12.1)–(12.3) indicated that long-chain fatty acids contributed 54 % of all acetyl-CoA entering the cycle (F_{C3}), lactate contributed 25 % (F_{C2}), and unlabeled substrates contributed 21 % (F_{C0}) in control hearts while the lactate contribution increased to 50 % and fatty acid contribution decreased to 20 % in hearts from hypertensive animals. The unlabeled contribution (probably mostly glucose) also increased in hypertensive hearts by about 9 %, consistent with an overall increase in carbohydrate oxidation in hearts from SHR animals. This direct analysis of the C4 resonance requires only that the TCA cycle intermediate pools have 'turned over' enough times after exposure to this ^{13}C-enriched substrate mix to enrich glutamate C3 sufficiently so that its resonance integral can be measured accurately and that the C4D34 and C4Q multiplets are visible in the spectrum. A more complete isotopomer analysis could be done by analysis of all multiplets in glutamate C2, C3 and C4, but the one presented above is simple and provides a quantitative measure of the metabolism of interest in this case that hearts from SHR animals oxidize significantly less fat and more carbohydrates than do hearts from control animals.

12.5. ISOTOPOMER ANALYSIS IN GLUCONEOGENIC TISSUES

Flux through anaplerotic pathways is relatively modest in non-gluconeogenic tissues such as heart so there is little dilution of label in TCA cycle intermediates on each turn of the cycle and the fractional ^{13}C enrichment of glutamate C2, C3 and C4 typically approaches the fractional ^{13}C enrichment in the acetyl-CoA pool. However, in gluconeogenic tissues such as liver, total anaplerotic influx can be quite high and, if the anaplerotic substrate is not enriched with ^{13}C, then the TCA cycle intermediates are continuously diluted by entry and mixing of natural abundance carbons. This has a dramatic effect on the appearance of the ^{13}C NMR spectrum of glutamate and all other associated intermediates. This is illustrated in Figure 12.5 for a tissue oxidizing [U-^{13}C]fatty acids. If anaplerotic flux is zero, the glutamate C4 resonance shows a substantial quartet (C4Q) that reflects a large population of isotopomers with ^{13}C enrichment of C3, C4 and C5. If on the other hand unlabeled anaplerotic substrate entered the TCA cycle at a rate equal to

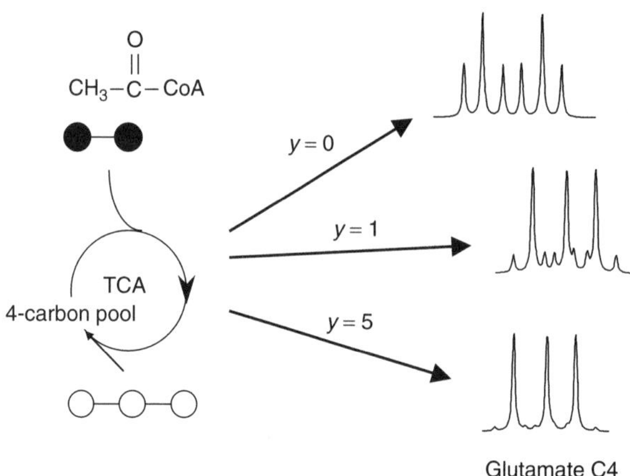

Figure 12.5. Schematic illustration of how the multiplets in glutamate C4 are diluted in highly anaplerotic tissues by entry of unlabeled substrate into the 4-carbon pools on each turn of the cycle. In this example, ^{13}C enters the cycle only from a substrate contributing to acetyl-CoA. The loss of multiplet information in the ^{13}C NMR spectrum makes an isotopomer analysis of glutamate problematical.

cycle flux (defined as $y = 1$), then ^{13}C enrichment of C3 would be diluted considerably and the intensity of C4Q would be reduced accordingly. For values of y greater than 1, the multiplet information in the glutamate spectrum is degraded to the point where isotopomer analysis becomes problematical and even the simple analysis described by Equations (12.1)–(12.3) may no longer easily provide the desired information about substrate utilization.

A solution to this isotopomer dilution problem is to change the delivery route of ^{13}C into the TCA cycle away from acetyl-CoA to a uniformly ^{13}C-enriched anaplerotic substrate. Using this strategy, a uniformly enriched three-carbon fragment enters the TCA cycle pools perhaps even multiple times with each turn of the TCA cycle and the resulting ^{13}C–^{13}C spin–spin interactions remain mostly intact and the multiplets in ^{13}C NMR spectra are not diluted by cycle activity or increased anaplerotic flux. Figure 12.6 illustrates that

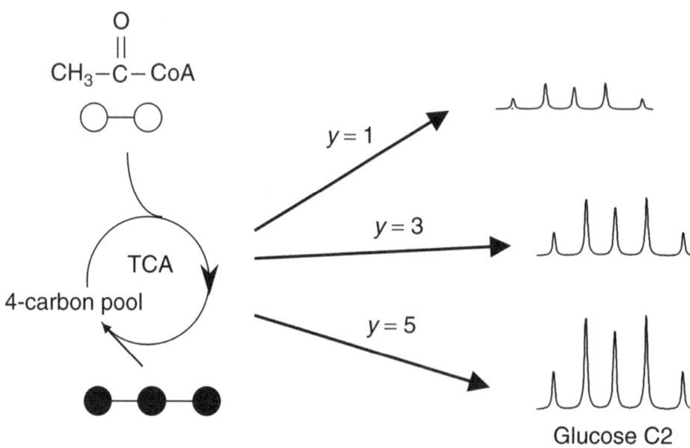

Figure 12.6. Schematic illustration of how ^{13}C multiplets are preserved in highly anaplerotic tissues by using a labeled anaplerotic substrate. In this example, ^{13}C enters the cycle only from a uniformly ^{13}C enriched three-carbon anaplerotic substrate. Although isotopomer information is lost in glutamate C4 and C5, it is preserved in other end-products of the cycle such as glucose.

the ^{13}C NMR spectrum of glucose derived from TCA cycle intermediates appears as multiplets regardless of anaplerotic flux, $y = 1$, 3 or 5. This is especially important for isotopomer analysis of the end-product of gluconeogenesis, glucose. The disadvantage of using a ^{13}C-enriched anaplerotic substrate instead of acetyl-CoA is that glutamate C4 and C5 no longer provide a direct readout of substrate preference as described above. A logical choice of labeled anaplerotic substrate for liver would be [U-^{13}C]pyuvate, [U-^{13}C]lactate or [U-^{13}C]alanine, typical gluconeogenic precursors. These tracers which would enter the cycle via pyruvate carboxylase and produce [1,2,3-^{13}C]oxaloacetate and, providing that equilibration through malate dehydrogenase and fumarate is fast, then this isotopomer also generates an equal amount of [2,3,4-^{13}C]oxaloacetate. Given that anaplerosis is typically $y \approx 3$ in liver, this makes it relatively easy to enrich OAA with a specific mix of isotopomers containing three enriched ^{13}C nuclei spin coupled to one another and, we shall see below, this mix of oxaloacetate (OAA) isotopomers is useful for unraveling some of the important exchanges that occur while converting OAA to glucose via gluconeogenesis. One disadvantage of [U-^{13}C]pyuvate, [U-^{13}C]lactate or [U-^{13}C]alanine as delivery vehicles that these same molecules can potentially contribute to acetyl-CoA via pyruvate dehydrogenase (PDH) and make the isotopomer analysis of glucose more complex. A convenient anaplerotic substrate that avoids this problem is [U-^{13}C]propionate, a three-carbon short-chain fatty acid normally present in the portal circulation of mammals. After activation to propionyl-CoA, this three-carbon unit is carboxylated to (*S*)-methylmalonyl-CoA, racemized to (*R*)-methylmalonyl-CoA, and finally converted to succinyl-CoA via the vitamin B_{12}-dependent enzyme, methylmalonyl-CoA mutase. This pathway is highly active in all mammalian tissues. Thus, [U-^{13}C]propionate directly produces [1,2,3-^{13}C]succinyl-CoA which enters the TCA cycle and is converted to equivalent amounts of [1,2,3-^{13}C]oxaloacetate and [2,3,4-^{13}C]oxaloacetate (Figure 12.7). This is the same result one would have obtained by using [U-^{13}C]pyuvate, [U-^{13}C]lactate or [U-^{13}C]alanine, but avoids the problems attendant with entry of these molecules into the oxidative pathways. Thus, [U-^{13}C]propionate is a convenient delivery vehicle for introducing a specific group of ^{13}C isotopomers into the TCA cycle. Furthermore, the cost of [U-^{13}C]propionate is not prohibitive and propionate can be delivered into the portal circulation by oral administration of sodium propionate in gel-caps in humans, or by i.p. injection in rodents. Propionate is also efficiently extracted by liver and used as a gluconeogenic precursor so one need not be concerned about extrahepatic metabolism of propionate. Finally, as illustrated in Figure 12.6, [U-^{13}C]propionate can be presented to tissues at tracer levels (1–2 % above natural abundance) and the

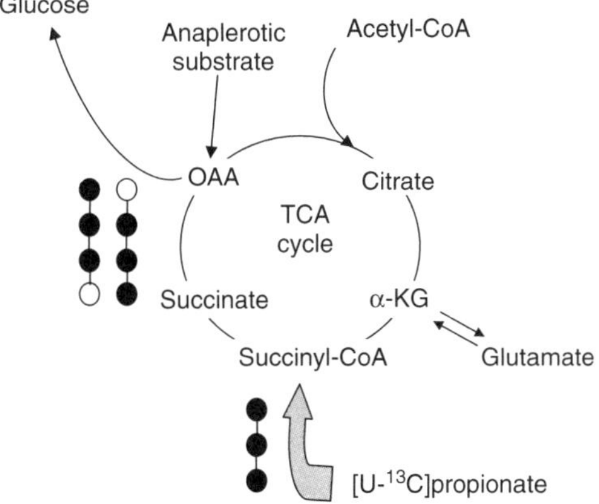

Figure 12.7. Schematic illustrating the utility of [U-^{13}C]propionate as a delivery vehicle of tracer for the TCA cycle. [U-^{13}C]propionate enters the cycle as [1,2,3-^{13}C]succinyl-CoA and is converted into equal populations of [1,2,3-^{13}C]oxaloacetate and [2,3,4-^{13}C]oxaloacetate.

metabolic information encoded in the ^{13}C spectrum as spin–spin multiplets is preserved in the ^{13}C NMR spectrum of glucose.

Using [U-^{13}C]propionate as a delivery vehicle of a specific groups of OAA isotopomers to the TCA cycle, it has been shown that flux through anaplerotic pathways, pyruvate cycling pathways and gluconeogenesis can be estimated from multiplet areas in the C2 resonance of glucose (or a derivative of glucose such as monoacetone glucose, MAG) all relative to TCA cycle flux (Jones *et al.*, 1997).

$$\text{Anaplerosis (OAA} \longrightarrow \text{PEP)} = (\text{D12} - \text{D23})/\text{D23} \tag{12.4}$$

$$\text{Pyruvate cycling (OAA} \longrightarrow \text{pyruvate} \longrightarrow \text{OAA)} = (\text{D12} - \text{Q})/\text{D23} \tag{12.5}$$

$$\text{Gluconeogenesis (PEP} \longrightarrow \text{glucose)} = (\text{Q} - \text{D23})/\text{D23} \tag{12.6}$$

These equations assume complete randomization of all symmetrical four carbon intermediates as [1,2,3-^{13}C]succinyl-CoA is converted to [1,2,3-^{13}C]oxaloacetate and [2,3,4-^{13}C]oxaloacetate, an assumption that is probably not valid for entry [U-^{13}C]lactate, [U-^{13}C]alanine or [U-^{13}C]pyruvate (via pyruvate carboxylase). There are certainly many examples in the literature of incomplete scrambling of OAA (derived from a labeled pyruvate) through fumarase prior to leaving the cycle as PEP destined for glucose. If 'backward' scrambling of OAA through fumarase is incomplete, then pyruvate cycling flux [Equation (12.5)] is underestimated.

12.6. ADDITION OF ^{2}H NMR DATA: IDENTIFYING THE SOURCES OF LIVER GLUCOSE

In an early study of metabolism in hepatocytes using ^3H_2O, it was concluded that the distribution of tritium in glucose could be used to assess gluconeogenesis (Rognstad *et al.*, 1974). Later, Kuwajima *et al.* (1986) injected tritiated water into experimental animals and used the distribution of ^{3}H in glucose to measure the relative flux through gluconeogenesis *versus* glycogenolysis. More recently, Landau *et al.* (1995, 1996) introduced ^2H_2O as a tracer for determining the sources of carbon for glucose production *in vivo*. Deuterated water has many advantages for *in vivo* studies: it is safe when taken orally by humans up to 0.3–0.5 % total body water (3–5 % in animals), it is relatively inexpensive, and it is easily administered and generally well-tolerated if spread over 2 h or more (Taylor *et al.*, 1966; Peng *et al.*, 1972; Money and Myles, 1974), even for infants (Sunehag *et al.*, 1999). After ingestion of ^2H_2O by fasting humans, glucose becomes enriched with ^{2}H at predictable positions within the glucose molecule, depending upon the pathway that generated the carbon fragment. Enrichment of the hydrogen attached to carbon 2 of glucose (H2) occurs at the level of glucose-6-phosphate (G6P) isomerase. Since this catalytic step is fast and considered near equilibrium, the products, G6P and F6P, are thought to be fully equilibrated under most circumstances. Thus, G6P derived from glycogen and destined for plasma glucose is thought to fully equilibrate through the isomerase before leaving the cell. Evidence for this comes from observations that ^{2}H enrichment at glucose H2 produced by liver equals total body water ^{2}H enrichment (Landau *et al.*, 1995). ^{2}H enrichment at H5, however, occurs at the level of the triose phosphates and therefore represents contributions from both PEP and glycerol (but not glycogen) to hepatic glucose output. Enrichment at one of the prochiral H6 hydrogens ($H6_S$) can be traced to the level of fumarase in the TCA cycle, so ^{2}H enrichment at this position reflects only the TCA cycle contribution to glucose production. It is then straightforward to determine the fraction of hepatic glucose released from each source provided that the ^{2}H enrichment in positions H2, H5 and $H6_S$ can be measured.

Two methods are currently available to measure ^{2}H enrichment in each site, mass spectrometry and ^{2}H NMR. Mass spectrometry requires sequential degradation of the glucose skeleton to generate formaldehyde

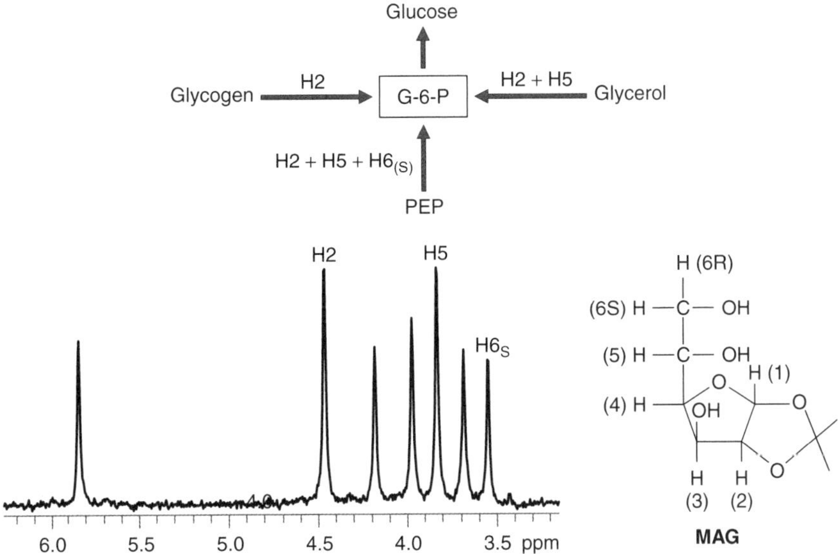

Figure 12.8. A ^{2}H NMR spectrum of MAG derived from plasma glucose of a rat after a 24 h fast. Plasma water contained ~3 % ^2H_2O. The areas of the H2, H5 and $H6_S$ resonances provide a direct measure of glycogen, glycerol and PEP to glucose production.

from each individual carbon which is then converted to hexamethylenetetramine (HMT) for mass spectral analysis (Landau *et al.*, 1996). Mass spectroscopy is also limited to one tracer per experiment, either ^{2}H or ^{13}C but not both. ^{2}H NMR spectroscopy is considerably less sensitive than mass spectroscopy but has three distinct advantages over mass spectrometry. First, a ^{2}H NMR spectrum provides fractional enrichments at all sites in a molecule in a single experiment and, second, the presence of ^{13}C at tracer levels does not interfere with the ^{2}H measurement. A third advantage is that the contributions of glycerol vs phosphoenolpyruvate (usually lumped together as 'gluconeogenisis') can be quantified separately in a single experiment. Glucose itself is not an attractive molecule for ^{2}H NMR because it has poor chemical shift dispersion and the spectrum is further complicated by the presence of α and β anomers (Jones *et al.*, 2000). However, chemical conversion of glucose to MAG (see Figure 12.8) removes the α and β anomers and yields a compound with seven well-resolved ^{2}H resonances (Jones *et al.*, 2001). A typical ^{2}H NMR spectrum of MAG derived from plasma glucose from a single fasted rat is shown in Figure 12.8. The relative ^{2}H enrichments at H2, H5 and $H6_S$ provide a direct readout of the three possible sources of glucose using the following relationships (Jones *et al.*, 2001).

$$\text{Glucose from PEP (Krebs cycle sources)} = H6_S/H2 \tag{12.7}$$

$$\text{Glucose from glycerol} = (H5 - H6_S)/H2 \tag{12.8}$$

$$\text{Glucose from glycogen} = 1 - (H5/H2) \tag{12.9}$$

The ^{2}H data can then be combined with the ^{13}C isotopomer measurement [Equations (12.4)–(12.6)] from the same sample to provide a rather complete metabolic map of liver metabolism from the TCA cycle to serum glucose *in vivo* (Figure 12.9). This combined ^{2}H and ^{13}C tracer experiment has been used to derive metabolic maps of metabolism in mice (She *et al.*, 2003), rats (Jones *et al.*, 1997, 1998a; Jin, *et al.*, 2003) and humans (Jones *et al.*, 1998b, 2001; Burgess *et al.*, 2003; Perdigoto *et al.*, 2003; Weis *et al*, 2004).

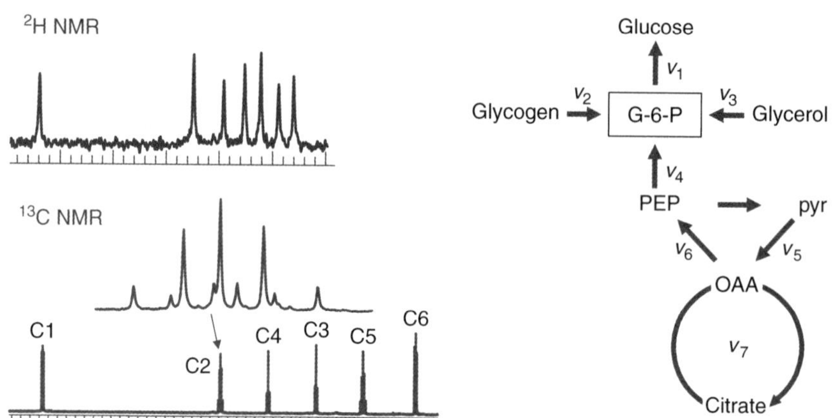

Figure 12.9. An illustration of how a single ^{2}H NMR spectrum and a single ^{13}C NMR spectrum of MAG links glucose production to TCA cycle flux. The ^{2}H data provides a readout of v_2, v_3 and v_4 all relative to v_1, while the ^{13}C data provides a readout of v_4, v_5 and v_6 all relative to v_7. Given a separate measure of glucose production (v_1), all remaining fluxes may be linked to TCA cycle flux (v_7).

12.7. THE GLUCONEOGENIC NETWORK: CONVERSION OF ^{2}H AND ^{13}C NMR DATA TO ABSOLUTE FLUXES

Given that ^{2}H NMR data report relative contributions of glycogen, glycerol and the TCA cycle to overall glucose production while the ^{13}C NMR data report anaplerosis, gluconeogenesis and pyruvate cycling all relative to TCA cycle flux, one can then combine these measurements to link glucose production with TCA cycle flux (Figure 12.9). Hepatic glucose production (HPG) can be measured in a separate experiment using any number of either radioisotope or stable isotope tracer molecules. A convenient tracer of HGP that does not interfere with the use of [U-^{13}C]propionate and ^2H_2O as tracers is [3,4-^{13}C]glucose (Jin *et al.*, 2004). Unlike other ^{13}C-enriched glucose tracers such as [1,6-^{13}C]- or [U-^{13}C]glucose, [3,4-^{13}C]glucose offers a distinct advantage in that ^{13}C from the tracer never enriches positions 1, 2, 5 or 6 of plasma glucose, even if the tracer is scrambled in the TCA cycle and glucose is eventually resynthesized. After infusion of [3,4-^{13}C]glucose at a constant known rate for period of time, plasma glucose can be isolated and converted to MAG and the residual doublets in the C3 and C4 resonances of MAG due to J_{34} coupling provide a direct measure of the amount of residual [3,4-^{13}C]glucose remaining at the end of the infusion period (Jin *et al.*, 2004). This provides a measure of HGP using standard assumptions. This absolute flux may then be tied to all relative flux values using:

$$v_1 = \text{hepatic glucose production } (\mu\text{mol/kg/ min}) \tag{12.10}$$

$$v_2 = \text{flux from glycogen} = v_1 \cdot (\text{H2} - \text{H5})/\text{H2} \tag{12.11}$$

$$v_3 = \text{flux from glycerol} = 2 \cdot v_1 \cdot (\text{H5} - \text{H6}_\text{S})/\text{H2} \tag{12.12}$$

$$v_4 = \text{flux from PEP} = 2 \cdot v_1 \cdot \text{H6}_\text{S}/\text{H2} \tag{12.13}$$

$$v_5 = \text{pyruvate cycling} = v_4 \cdot (\text{D12} - \text{Q})/(\text{Q} - \text{D23}) \tag{12.14}$$

$$v_6 = \text{flux through PEP carboxykinase} = v_4 \cdot (\text{D12} - \text{D23})/(\text{Q} - \text{D23}) \tag{12.15}$$

$$v_7 = \text{TCA cycle flux} = v_4 \cdot (\text{D23})/(\text{Q} - \text{D23}) \tag{12.16}$$

This simple set of equations summarizes how seven absolute fluxes ($v_1 - v_7$) can be derived from two NMR spectra (one ^{2}H and one ^{13}C) of a single derivative of plasma glucose (MAG). Although the method does require purification of plasma glucose and chemical conversion to MAG, both processes are amenable to robotic manipulation and the NMR data can be typically collected in a few hours using a high field NMR system (400–600 MHz). This makes the entire procedure attractive for high-throughput analysis of metabolic flux for screening drugs, for phenotyping transgenic animals, or for studying large patient populations of human disease.

12.8. ONE INTERESTING EXAMPLE: THE CASE OF LIVER-SPECIFIC PEPCK KNOCKOUT MOUSE

An interesting example of the metabolic consequences of a liver-specific knockout of PEP carboxykinase was recently reported by She *et al.* (2003). This enzyme catalyzes the first step of gluconeogenesis from the TCA cycle, the conversion of OAA to PEP. One would predict that mice lacking this liver enzyme would be severely hypoglycemic and perhaps not survive long after birth. Surprisingly, the liver-specific PEPCK knockout mouse had normal levels of plasma glucose and substantially less liver glycogen than controls, yet the knockout mice were able to withstand a 24 h fast. Radiotracer experiments in livers perfused *in situ* verified that they were unable make glucose from a mixture of lactate/pyruvate but had an increased capacity to make glucose from glycerol. This interesting mouse offered an opportunity to test the ^{2}H and ^{13}C tracer methods described above. Accordingly, livers from 24 h fasted PEPCK knockout mice were isolated and perfused with a mixture of unenriched glycerol (to provide a gluconeogenic precursor) and octanoate (to provide acetyl-CoA units for the TCA cycle), [U-^{13}C]propionate as a tracer of TCA cycle activity and ^2H_2O. Glucose production from this substrate mix was lower in livers from PEPCK knockout mice compared with livers from littermate controls by 50–100 %, depending upon fasting state of the animal. Glucose produced by these livers was isolated, converted to MAG and examined by ^{2}H and ^{13}C NMR and, as expected, $H6_S$ had only trace levels of ^{2}H, consistent with a lack of gluconeogenesis from TCA cycle precursors (Figure 12.10). The near equivalence of the H2 and H5 resonances in the ^{2}H spectrum indicates that essentially all of the glucose produced by this liver came from glycerol, and

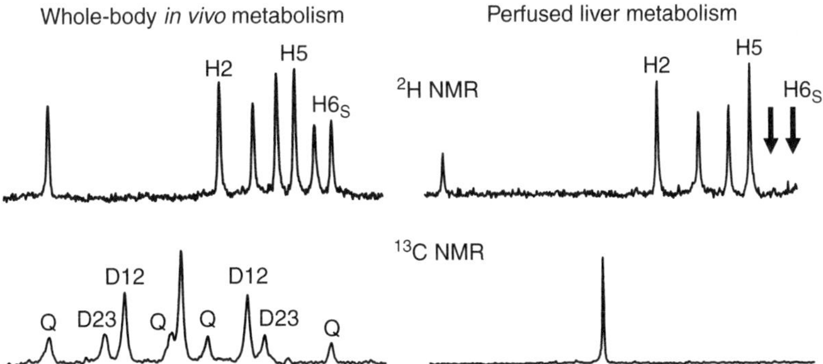

Figure 12.10. ^{2}H NMR and ^{13}C NMR of MAG from liver-specific PEPCK knockout mice. The left panel shows spectra of MAG from plasma glucose of the whole animal after i.p. injection of [U-^{13}C]propionate and ^2H_2O while the right panel shows spectra of MAG from glucose produced by isolated livers from the same animals after perfusion with [U-^{13}C]propionate, ^2H_2O, unenriched octanoate, lactate, pyruvate and glycerol. While gluconeogenesis from the TCA cycle is active in whole body metabolism, the isolated liver cannot produce glucose from TCA cycle precursors [as indicated by the absence of labeling in H6 (arrows) and the absence of multiplets in the ^{13}C NMR spectrum].

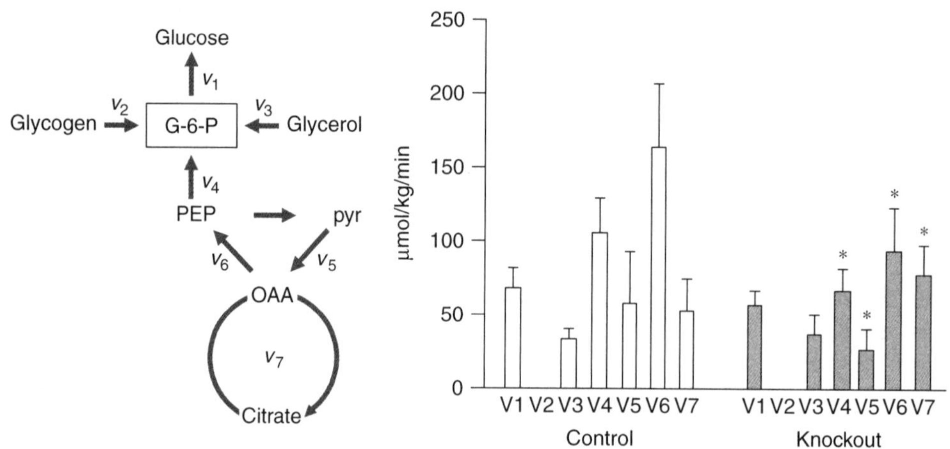

Figure 12.11. Summary of whole body fluxes (μmol/kg/min) in control (white bars) and PEPCK knockout animals (colored bars). These data were derived from a measure of total body glucose turnover, the peak areas of H2, H5 and $H6_S$ in ^2H NMR spectra and from multiplet areas in the C2 resonance of MAG in ^{13}C NMR spectra ($*p \leq 0.05$ vs controls, $n = 4$).

the ^{13}C NMR spectrum of MAG verified that plasma glucose had very little ^{13}C enrichment from [U-^{13}C]propionate. This experiment proved that both tracer molecules (^2H_2O and [U-^{13}C]propionate) correctly reported the expected result.

Even more interesting were the results of *in vivo* experiments with these PEPCK knockout mice (She *et al.*, 2003). After giving ^2H_2O and [U-^{13}C]propionate as a single i.p. injection, plasma glucose was isolated, converted to MAG and analyzed by ^2H and ^{13}C NMR. In this case, the ^2H spectrum of the KO animal (Figure 12.10) was nearly identical to a ^2H spectrum from a littermate control (not shown). Glucose turnover as measured using a [^3H]glucose tracer was slightly lower in the KO animal compared with control, and the multiplets in the C2 resonance of MAG showed a slight increase in D23 and decrease in Q in samples from knockout animals. These relatively minor differences, however, translated into significant differences (Figure 12.11) in total gluconeogenic flux (lower in knockout animals), total anaplerosis (lower in knockout animals), pyruvate cycling (lower in knockout animals) and TCA cycle flux (higher in knockout animals). Given that glucose cannot be made from the TCA cycle in PEPCK knockout animals, the observation that $H6_S$ is labeled in glucose from the knockout animals means that it originated in an organ other than liver. This also indicates that the NMR flux values in this case represent an average of all glucose production likely from multiple tissue sources. It is interesting to speculate, however, that the higher TCA cycle flux measured in the knockout animals indicates that the TCA cycle of those tissues making glucose in these animals expend more energy on other biological processes than does the liver in control littermates. This could reflect the high metabolic activity of kidney, for example, but this has not been confirmed.

12.9. FUTURE DIRECTIONS: COMPREHENSIVE, HIGH-THROUGHPUT METABOLIC ANALYSES

Much of this chapter has focused on NMR tools for easily identifying the sources of glucose production (glycogen vs glycerol vs TCA cycle precursors) and the link between glucose output and metabolic fluxes associated with the TCA cycle. While fluxes in many of the key pathways involved in glucose production have been described in isolated cell and organs, there is considerable uncertainty regarding the relevance of these measurements to intact organisms and particularly in humans with diseases where cellular organization, substrate availability and other factors may not be duplicated in simpler systems. The availability of convenient, non-invasive methods to quantify these fluxes in humans may prove important in understanding the pathogenesis of disease and perhaps aid in therapeutic decisions. Type 2 diabetes, for example, is a heterogeneous disorder with enormous morbidity and multiple therapeutic options. It would seem reasonable in the twenty-first century that drug therapy could be tailored to the individual, based on the known effects of the agents and determination of the relative importance of gluconeogenesis, glycogenolysis, hepatic insulin resistance or skeletal muscle insulin resistance in the maintenance of hyperglycemia. Unfortunately, that is not the case. Stratification of patients with type 2 diabetes or other metabolic derangements according to metabolic fluxes may ultimately prove useful in understanding genetic and environmental factors that predispose obesity and its related disorders of glucose metabolism. NMR and stable isotope tracers will probably play an important role in helping to unravel these complex metabolic questions.

Acknowledgments

This work was supported by the National Institutes of Health (HL-34557 and RR-02584), the National Science Foundation, the Department of Veterans Affairs, the University of Texas at Dallas, and the University of Texas Southwestern Medical Center. The authors thank Shawn Burgess and Eunsook Jin for allowing us to describe some of their data prior to publication.

REFERENCES

Burgess SC, Weis B, Jones, JG, Smith E, Merritt ME, Margolis D, Sherry AD, Malloy CR. Noninvasive evaluation of liver metabolism by ^{2}H and ^{13}C NMR isotopomer analysis of human urine. *Analytical Biochemistry* 2003; **312**: 228–234.

Chance EM, Seeholzer SH, Kobayashi K, Williamson JR. Mathematical analysis of isotope labeling in the citric acid cycle with applications to ^{13}C NMR studies in perfused rat hearts. *Journal of Biological Chemistry* 1983; **258**: 13785.

Eakin RT, Morgan LO, Gregg CT, Matwiyoff NA. Carbon-13 nuclear magnetic resonance spectroscopy of living cells and their metabolism of a specifically labeled ^{13}C substrate. *FEBS Letters* 1972; **28**: 259.

Jin ES, Uyeda K, Kawaguchi T, Burgess SC, Malloy CR, Sherry AD. Increased hepatic fructose 2,6-bisphosphate after an oral glucose load does not affect gluconeogenesis. *Journal of Biological Chemistry* 2003; **278**: 28427–28433.

Jin ES, Jones JG, Merritt M, Burgess SC, Malloy CR, Sherry AD. Glucose production, gluconeogenesis, and hepatic TCA cycle fluxes measured by NMR analysis of a single glucose derivative. *Analytical Biochemistry* 2004; **327**: 149–155.

Jones JG, Naidoo R, Sherry AD, Jeffrey FM, Cottam GL, Malloy CR. Measurement of gluconeogenesis and pyruvate recycling in the rat liver: a simple analysis of glucose and glucamate isotopomer during metabolisms of [1,2,3-$^{13}C_3$] propionate. *FEBS Letters* 1997; **412**: 131–137.

Jones JG, Carvalho RA, Franco B, Sherry AD, Malloy CR. Measurement of hepatic glucose output, Krebs cycle and gluconeogenic fluxes by NMR analysis of a single plasma glucose sample. *Analytical Biochemistry* 1998a; **263**: 39–45.

Jones JG, Solomon MA, Sherry AD, Jeffrey FM, Malloy CR. ^{13}C NMR measurements of human gluconeogenic fluxes after ingestion of [U-^{13}C] propionate phenylacaetate and acetaminophen. *American Journal of Physiology* 1998b; **275**: E843–E852.

Jones JG, Carvalho RA, Sherry AD, Malloy CR. Quantitation of gluconeogenesis by ^{2}H nuclear magnetic resonance analysis of plasma glucose following ingestion of $^{2}H_2O$. *Analytical Biochemistry* 2000; **277**: 121–126.

Jones JG, Solomon MA, Cole SM, Sherry AD, Malloy CR. An integrated ^{2}H and ^{13}C NMR study of gluconeogenesis and TCA cycle flux in humans. *American Journal of Physiology* 2001; **281**: E848–E856.

Jucker BM, Lee JY, Shulman RG. *In vivo* ^{13}C NMR measurements of hepatocellular tricarboxylic acid cycle flux. *Journal of Biological Chemistry* 1998; **273**: 12187.

Kuwajima M, Golden S, Katz J, Unger RH, Foster DW, McGarry JD. Active hepatic glycogen synthesis from gluconeogenic precursors despite high tissue levels of fructose 2,6-bisphosphate. *Journal of Biological Chemistry* 1986; **261**: 2632–2637.

Landau BR, Wahren J, Chandramouli V, Schumann WC, Ekberg K, Kalhan SC. Use of $^{2}H_2O$ for estimating rates of gluconeogenesis: application to the fasted state. *Journal of Clinical Investigation* 1995; **95**: 172–178.

Landau BR, Wahren J, Chandramouli V, Schumann WC, Ekberg K, Kalhan SC. Contributions of gluconeogenesis to glucose production in the fasted state. *Journal of Clinical Investigation* 1996; **98**: 378–385.

London RE. ^{13}C labeling in studies of metabolic regulation. *Progress in NMR Spectroscopy* 1988; **20**: 337.

Malloy CR, Sherry AD, Jeffrey FM. Carbon flux through citric acid cycle pathways in perfused heart by ^{13}C NMR spectroscopy. *FEBS Letters* 1987; **212**: 58–62.

Malloy CR, Sherry AD, Jeffrey FM. Evaluation of carbon flux and substrate selection through alternate pathways involving the citric acid cycle of the heart by ^{13}C NMR spectroscopy. *Journal of Biological Chemistry* 1988; **263**: 6964–6971.

Malloy CR, Sherry AD, Jeffrey FM. Analysis of tricarboxylic acid cycle of the heart using ^{13}C isotope isomers. *American Journal of Physiology* 1990a; **259**: H987–H995.

Malloy CR, Thompson JR, Jeffrey FM, Sherry AD. Contribution of exogenous substrates to acetyl coenzyme A measurement by ^{13}C NMR under non-steady-state conditions. *Biochemistry* 1990b; **29**: 6756–6761.

Money KE, Myles WS. Heavy water nystagmus and effects of alcohol. *Nature* 1974; **247**: 404–405.

Peng SK, Ho KJ, Taylor CB. Biological effects of prolonged exposure to deuterium oxide. *Archives of Pathology* 1972; **94**: 81–89.

Perdigoto R. Furtado AL. Porto A. Rodrigues TB. Geraldes CF. Jones JG. Sources of glucose production in cirrhosis by $^{2}H_2O$ ingestion and ^{2}H NMR analysis of plasma glucose. *Biochimica et Biophysica Acta* 2003; **1637**: 156–163.

Rognstad R, Clark G, Katz J. Glucose synthesis in tritiated water. *European Journal of Biochemistry* 1974; **47**: 383–388.

Rothman DL. Magnusson I. Katz LD. Shulman RG. Shulman GI. Quantitation of hepatic glycogenolysis and gluconeogenesis in fasting humans with ^{13}C NMR. *Science* 1991; **254**: 573–576.

She P, Burgess SC, Shiota M, Flakoll P, Donahue EP, Malloy CR, Sherry AD, Magnuson MA. Mechanisms by which liver-specific PEPCK knockout mice preserve euglycemia during starvation. *Diabetes* 2003; **52**: 1649–1654.

Sherry AD, Jeffrey FM, Malloy CR. Analytical solutions for ^{13}C Isotopomer analysis of complex metabolic conditions: substrate oxidation, multiple pyruvate cycles and gluconeogenesis. *Metabolic Engineering* 2004; **6**: 12–24.

Sunehag AL, Haymond MW, Schanler RJ, Reeds PJ, Bier DM. Gluconeogenesis in very low birth weight infants receiving total parenteral nutrition. *Diabetes* 1999; **48**: 791–800.

Taylor CB, Mikkelson B, Anderson JA, Forman DT. Human serum cholesterol synthesis measured with the deuterium label. *Archives of Pathology* 1966; **81**: 213–232.

Weis BC, Margolis D, Burgess SC, Merritt ME, Wise H, Sherry AD, Malloy CR. Glucose production pathways by ^{2}H and ^{13}C NMR in patients with HIV-associated lipoatrophy. *Magnetic Resonance in Medicine* 2004; **51**: 649–654.

13

Summarized Reflections on Metabolism

Robert G. Shulman, Douglas L. Rothman and James R. A. Schafer

Departments of Diagnostic Radiology and Neuroscience, Yale University School of Medicine, MR Center, New Haven, CT 00520-8043, USA

Recent progress in molecular biology and biochemistry has become overwhelming with its unrelenting revelations about the complexity of the living state. Advances at the molecular level have unveiled great libraries of nature's secrets – with data collections so large that new technologies have had to be developed just to store them. Molecular genetics, structural biology and intermediary metabolism have identified nucleic acids and proteins involved in an ever-expanding network of biochemical activities. The details of molecular properties are so plentiful that they can only be handled by computers. The earlier wall charts of intermediary metabolism whose intricacies could dazzle visitors now seem quaint relics of a far simpler time. Banks of data being collected by genomics and proteomics await identification and explanation. Life, it is now being proposed, consists of sets of data described in awesome detail merely awaiting interpretation.

Metabolomics by In Vivo NMR. Edited by R. G. Shulman and D. L. Rothman
 ISBN: 0-470-84719-0

Analysis of a complex phenomenon in terms of its components is a methodology responsible for advances in the physical sciences since Descartes' time. The difficulty in applying this method to the study of life is in turning the reductionist understanding back upon the more complex. The powers of modern structural and molecular biology have resulted in so much detail at the component level that efforts to integrate the data with higher organizational levels devolve into hopes for super-molecules (i.e. everything we need to know about Alzheimer's disease can be found in amyloid precursor protein) or vague assertions that faster computers and bigger data sets will magically resolve the problem at some future time. The integration problem is made still worse by the parallel increase in our understanding of the more complex aspects of the life sciences. Organism-level biological phenomena, such as disease and behavior, have been described much more completely and yet, despite much study, still mostly lack mechanistic explanations. As a consequence, modern biochemistry sits at the juncture of two universes – that of complex organism-level properties and that of the equally complex and ever more precisely described molecular properties – which appear to be drifting further and further apart.

To understand the biological functions of genes and proteins (i.e. to bridge the gap from top down) and to base understanding of the complex organism upon molecular findings (i.e. bottom up), it is necessary to develop methods that study the living state at some intermediate, linking, level of complexity. This understanding would find mechanisms that integrate functions regardless of level. Although many directions are being explored, in this book we have focused on recent advances made in the oft-neglected field of intermediary metabolism – a science which functions at a level one step up from biochemical molecules. It is in metabolic pathways that protein molecules, produced by the genes and with properties discernable by physics and chemistry, coordinate to serve physiological and organismic needs. At this level physiological and cellular activities, based upon molecules, respond both to external conditions and to the intrinsic needs of a cell. Intermediary metabolism, and the closely related field of systemic physiology, was considered by many to be static, no longer an exciting research subject compared with the recent vigorous advances in molecular and structural biology. While its location on the epistemological ladder between molecular biochemistry and systemic physiology has made it too important to be overlooked, it has frequently been regarded as a textbook subject with neatly stacked results, to be used whenever necessary. We propose that it is time to dramatically revise this view. The dynamic new physical methods that are discussed in this book not only illustrate and increase the importance of metabolism, but also show how new concepts of intermediary metabolism play a crucial role in linking biological levels. We propose that metabolism provides a key bridge that can connect complex phenomenon at the physiological and behavioral level with molecular mechanisms of gene expression and protein function.

Since 1981, the challenge of studying biochemistry in the living state has been addressed by developing ^{13}C nuclear magnetic resonance spectroscopy (MRS) for measuring concentrations of metabolites and the rates through metabolic pathways in the human body. *In vivo* NMR spectroscopy identifies the resolved NMR signals from different chemicals in the body, a method that is well developed in organic chemistry. As a result, *in vivo* NMR can measure the concentrations and synthesis rates of individual biological molecules such as glycogen, glycolytic intermediates and neurotransmitters, within precisely defined areas of specific organs such as brain, liver and muscle. Thus, it has the unique capacity to reveal how critical metabolic pathways actually work in healthy individuals and how they are altered in disease.

The use of ^{13}C NMR to track metabolic pathways *in vivo* was pioneered in the 1970s. Metabolic substrates such as glucose were tagged with an NMR-visible stable isotope, ^{13}C, and *in vivo* NMR measurements then followed the metabolism of the labeled glucose through the appearance of enriched ^{13}C isotope in metabolites. Rates of these metabolic pathways were determined by measuring the label appearance as a function of time. Initial studies were performed with microorganisms and perfused organs. These experiments were extended to intact animal models in 1981 and soon afterwards to human subjects when superconducting magnets capable of following *in vivo* NMR spectroscopy in humans were developed. Since that time, *in vivo* ^{13}C NMR has been applied to study a variety of metabolic pathways and systems in animals and humans in our laboratory and in other laboratories worldwide.

The meaningfulness of *in vivo* MRS results is reinforced and extended by metabolic control analysis (MCA). MCA offers an opportunity to replace the chaotic contradictory descriptions of metabolic regulation and control with a formal structure that is logical, quantitative and experimentally accessible. It provides a model for understanding metabolic results that is analogous to the role that Michaelis–Menten kinetics played in enzymatic kinetics. Once the conditions assumed in its derivation are met, MCA allows experimental results to be analyzed and integrated very much as enzymes are described by their K_m and V_{max} values. Much of the development of MCA has focused on facilitating the use of experimental data so as to relate the regulation and control of metabolism to properties of the constituent enzymes, hormones and metabolites. Since MCA shows the dependence of control not only on these components but also upon the external conditions, it is ideally suited for explaining control *in vivo*, providing that *in vivo* data are available. It is here that it integrates with MRS, which can evaluate rates of metabolic pathway and concentrations of metabolites under specific conditions *in vivo*. Finally, as discussed below, understanding how biochemical pathways *in vivo* are satisfying physiological conditions begins to integrate the molecular and systemic properties – the challenge of modern biochemistry.

In this book we have focused on studies performed primarily in the muscle with less detailed studies of heart, liver and yeast. It complements the previous volume *NMR Studies of Brain In Vivo: Application to fMRI and Mind*. Both have probed cellular metabolism in the living human and have provided new understanding of metabolism as a link between molecular- and organism-level functioning. Non-invasive NMR studies of metabolism are at the forefront of a reinvigorated science of intermediary metabolism poised to make enormous contributions to biology by connecting biological elements that have been allowed to drift too far apart.

13.1. METABOLISM OCCUPIES A CENTRAL POSITION IN LIFE, LINKING GENES AND PROTEINS TO HIGHER-LEVEL BIOLOGICAL FUNCTIONS

A metabolic pathway is a series of chemical reactions that the cell uses to break down simple molecules such as glucose for fuel, or to synthesize more complex biomolecules such as hormones and neurotransmitters. The ability of the body to perform the integrated physiological processes needed for survival, such as providing the energy needed to sustain brain function, depends upon the coordination of metabolic pathways in billions of cells. Metabolic pathways thus occupy a central position in biology – extending from the reductionist end with genes and the proteins they express in the individual cell upward to complex interactions in physiology, organ function and behavior at the level of the organism.

The research strategy we advocate is to work from metabolism in a step-by-step manner, developing mechanisms that reach both upwards in complexity, to understand how metabolism supports higher level functions, and downwards towards a better physical understanding of the roles of specific genes and proteins in controlling metabolism. Of foremost importance are methods that have been developed for determining the roles played by enzyme kinetics in metabolism and for understanding the changes in metabolism in response to the perturbations required by systemic physiology. Prominent amongst these methods is MCA, which evaluates experimental data quantitatively and can guide the design of experiments. A key insight of MCA is that the degree of control an enzyme exerts on a metabolic pathway is not constant but rather depends upon the environment of the cell, the expression of other enzymes in the pathway, and the activity of other metabolic pathways in the cell. Therefore it emphasizes the critical need to study metabolism *in vivo* in order to understand the function of specific enzymes, and their regulation through gene expression and cellular signaling. These findings make it unlikely that the functions of proteins in an organism will be understood merely through correlations of gene alterations with complex phenotypical traits. Only a more systemic approach will ultimately provide a satisfactory account of protein function *in vivo.*

13.2. NATURE AND NURTURE IN NON-INSULIN-DEPENDENT DIABETES MELLITUS

A great recurring unsolved biological problem that genomics, with its implicit claims of genetic determinism, has once again brought to the fore, centers on the roles of genetics and environment in determining human biology; the long-standing nature vs nurture dialog. Recent biological theorists, particularly Richard Lewontin, have suggested that this dialog is more realistically a three-party interaction because these two classical forces are under the influence of chance events which meet and are expressed in an individual organism, itself the result of an undetermined history. To seek the primacy of genetic or environmental forces, statistical analyses of populations are often conducted so as to assess causal relationships of these variables. As an alternative we propose that studying metabolic consequences of genetic and environmental factors illuminates their interplay in individual humans, and reduces the uncertainties introduced by contingency.

^{13}C MRS studies of intermediary metabolism in non-insulin dependent diabetes mellitus (NIDDM) patients have established the main relevant pathway of impaired glucose storage, and have explored effects of controlled environmental factors upon genetically altered pathways in the etiology of the disease. Glycogen metabolism in healthy human controls is compared with NIDDM patients and their offspring in Chapter 4. NIDDM has a genetic component, as shown by its high coincidence in identical twins and by numerous family histories. It was known before the MRS studies that in early life NIDDM is manifested by impaired glucose tolerance, which is compensated for by pancreatic up-regulation of insulin secretion – thus forestalling chronic hyperglycemia and its attendant symptomatology until pancreatic function decays, perhaps as a result of chronic insulin overproduction. Before 1990 the metabolic analysis of NIDDM suggested that liver and/or muscle were the leading contenders for the specific molecular site of the disease, with passionate advocacy for different enzymes and pathways at each site. In 1990 ^{13}C MRS identified the muscle pathway of glycogen synthesis as the main avenue of glucose storage in healthy controls. Further it showed that in NIDDM patients derangement of this pathway was responsible for impaired glucose tolerance and the subsequent hypoglycemia, as described in Chapter 4. MRS experiments, with interpretations guided by MCA, then showed that the flux of glycogen synthesis was controlled by the first step in the pathway, the recruitment of glucose transporters to the plasma membrane. This recruitment was identified as the metabolic site of the genetic lesions that could, under suitable conditions, lead to disease as well as the defect for which pancreatic over-production of insulin was compensating in the pre-disease state. The actual appearance and severity of the disease was well known from statistical population studies to depend upon parameters of lifestyle, predominantly diet, exercise and obesity. Management of these factors plays a major role in public health policies seeking to prevent or minimize NIDDM. Chapter 4 explores the influence of these environmental factors on the glycogen synthesis pathway in NIDDM patients. It is shown that the environment acts on subjects with a genetic susceptibility to NIDDM through the same molecular mechanisms. The difference in metabolic phenotype is due to enhanced sensitivity to environmental factors which reduce insulin sensitivity in pre-diabetics.

The ongoing gene vs environment debate about human diseases as played out in NIDDM subjects shows that, in certain well-studied cases, lifestyle (environmental) parameters (e.g. obesity and exercise) can be interpreted in terms of their physical-chemical effects upon the pathway identified as the locus of the mutation. This brings the individual's response to the center of the genetic/environmental debate, as Lewontin proposed. However instead of leaving the interaction completely to chance, as necessarily concluded by evolutionary biologists without the benefit of biochemical knowledge, studies of *in vivo* metabolism by MRS and MCA can, in well-studied cases such as NIDDM patients, set limits upon 'chance' by establishing biochemical causality.

13.3. PHOSPHORYLATION, MULTISITE MODULATION AND HOMEOSTASIS

A second revolutionary finding from the MRS studies of glycogen synthesis is the novel role found for enzymatic phosphorylation in a pathway. MCA defines flux control coefficients as the fractional change in flux divided by the fractional change in the enzyme's activity. Hence, when glycogen synthase (GSase) was shown to have a flux control coefficient value near zero for the glycogen synthesis pathway (Chapter 5), the question arose as to why the activity of this allosteric enzyme was controlled by an extensive array of second messengers, hormones, assorted kinases and phosphatases. The answer proposed in Chapter 5 was that, while pathway flux is controlled by glucose transporter regulation of plasma glucose movement into the cell, GSase serves to maintain constant concentrations of pathway metabolites by changing its activity to match transporter activity. Thus, GSase is controlled by phosphorylation so as to maintain homeostasis of the intracellular intermediates while the glycogen synthesis flux is controlled by glucose transporters so as to maintain homeostasis of plasma glucose.

These results support and extend the applicability of two powerful hypotheses that affect the control of metabolism. Although both had been suggested earlier in MCA studies, the experimental results in Chapter 5 and elsewhere in this book opened the road to their revolutionary consequences. The first direct consequence of the GSase findings is that, if phosphorylation/dephosphorylation of an enzyme occurs and changes the activity of the enzyme, it does not mean that the targeted enzyme is controlling the flux of its pathway. Like GSase, these enzymes may be maintaining homeostasis of the pathway intermediates while the control of flux is elsewhere in the pathway. The bewildering abundance of allosteric phosphorylated enzymes in many pathways might serve to maintain homeostasis of metabolite concentrations rather than to control fluxes. It had previously been proposed that homeostasis could be maintained by coordinated changes in concentrations of all enzymes in a pathway via coordinated changes in gene expression. The GSase results illustrate a case where homeostasis of intermediates is maintained by phosphorylation control of a single step. In a longer pathway (e.g. glucose oxidation through glycolysis and oxidative phosphorylation down to adenosine triphosphate, ATP), the high number of phosphorylated, allosteric enzymes could now be taken as localized maintainers of concentrations, providing a boost at each step, analogous to the functions of amplifiers at intervals in a long transmission line, that keep, in that case, the signal level close to constant.

The meaningfulness of these results increases when we consider their impact upon the next hypothesis that they generated. In the several common pathways studied in this book, MCA shows that flux control is exercised either by supply of substrates (e.g. glucose controls glycogenesis) or by the demand for a product which happens to be ATP (e.g. in muscle work and spore survival). Two of these pathways (e.g. glycogenesis and muscle work), have been studied over several decades and the location and mechanism of flux control have been intensively and actively contested. Even when there is a consensus on general energetics (e.g. that in muscle the rate of energy production is regulated to meet the demands of work) the proponents of flux control by PFK (pyruvate kinase), or by oxygen supply, or by cytochrome oxidase, have not abandoned their efforts to establish each of these steps as the controller.

At first it was accepted that MCA, with its definition of 'flux control coefficients' that could distribute control, had resolved the problem. With distributed flux control, it was argued, everyone could be correct – all these steps could be contributing. However, from the findings in this book an alternate paradigm must be considered. Although MCA shows how flux control could be widely distributed, it turns out in these three cases not to be. The flux of glycogenesis is controlled by the plasma glucose uptake step so as to maintain constant levels of glucose. Muscle and spore ATP production are controlled by the demand for ATP, not by steps in the production pathway. In these cases the biochemical pathway exists so that its flux serves the physiological needs for homeostasis of the supply substrates (e.g. glucose) or for the products (like ATP) demanded, while at the same time it maintains homeostasis of the biochemical intermediates so as not to disrupt other biochemical reactions. To the extent that the emphasis placed upon homeostasis by these selected results is widespread, we begin to envisage a new role for the constraints of biochemical pathways. In this view, biochemical fluxes are at the service of the body whose needs are expressed in physiological terms. The study of intermediary metabolism in the present cases reveals how the pathways serve the physiological needs of the organism. It remains to be seen how common the localization of flux control to the first or last step in a pathway is, but the large role of homeostasis in biochemical pathways has implications for biochemistry and physiology. The complex regulation by signaling of many of the proteins in the cell allows the cell to monitor homeostasis and maintain metabolic integrity despite large changes in pathway flux required by the organism. Hofmeyr and Cornish-Bowden while developing supply on demand control concepts from MCA had shown that the first or last enzymatic steps would have the greatest flux control, in accord with our findings.

13.4. BIOCHEMICAL HOMEOSTASIS AND SYSTEMS PHYSIOLOGY

Multisite control allows large flux changes to be accomplished within a cell with a minimum of disruption to internal homeostasis. One way in which this may be accomplished is through the direct modification by some external effector of the activity of the enzymes in question. While theoretically feasible, this arrangement would require this outside effector to exactly link the kinetics of all its targets in perfect synchrony with pathway input and output in order to prevent local disequilibriums with resultant dissipation/accumulation of metabolites. The glycogen synthase pathway provides examples of two strategies that address this problem. The first is the present of metabolically inactive 'sinks', such as glycogen, which allow the output to vary harmlessly. The second, and presumably more important, mechanism is the division of homeostatic and flux control between the modulated enzymes. An increase in insulin induces an increase in pathway flux by elevating the activity of GT. However, it also increases the sensitivity of GSase to an upstream intermediate – thus allowing the flux changes induced upstream to exert their effects with minimal disruption of homeostasis and with a minimum of precise downstream insulin flux control. Given the success of this system of proportional activation (Chapter 5) in rapidly moving excess substrate out of the circulation and in maintaining internal homeostasis, we imagine that analogous arrangements will be common.

Beyond the internal accommodations these findings offer a basis for speculation about coupling between different pathways in response to physiological needs. A state change, introduced for example by feeding or fasting, is met by widespread metabolic changes. How are these changes coordinated? If they depend upon coupled changes of flux control, the near infinite number of conditions and couplings is daunting since they depend upon the assembly of hormones, second messages and metabolites generated by the change of state. If on the other hand, flux control is mainly exercised in response to external effectors by a limited number of pathway enzymes, then the rest of the enzymes could serve homeostasis. We have also already seen in the glycogen synthesis pathway how homeostasis is accommodating, sometimes even slightly inaccurately in response to flux change, so that a hormone like insulin would not have to control flux changes in all of the many pathways it affects but it might contribute to homeostasis in the many pathways which contain its targeted enzymes. We see that proportional activation creates pathway homeostasis by multisite modulation effected by allosteric phosphorylated enzymes. This effectively brings a pathway to act as a single enzyme in response to physiological demands, allowing pathways to function efficiently together.

13.5. YEAST AND YEAST SPORES

The system of intermediary metabolism involves building blocks of metabolites, enzymes and macromolecular complexes with a high degree of conservation between even widely divergent species. For example, the glycolytic pathway is almost identical across phyla from bacteria to mammals. This consistency is so prevalent that biochemistry textbooks usually describe pathways without reference to their parent organism. Obviously, natural selection has found great utility in these ubiquitous pathways and they have become as much a part of terrestrial life as the genetic code.

However, just as that code is used in wildly different ways by different organisms, the catalytic steps of the glycolytic pathway serve different purposes in different organisms and therefore the metabolic control of the same pathway can be quite variable. For instance, while mammalian cells must concern themselves with maintaining homeostasis not just internally but in their aggregate organism, microorganisms such as yeast must respond as isolated units to dramatic changes in their environment. Their goals thus appear more self-centered: they need to germinate, to grow and to divide whenever conditions are favorable. They have to aggressively make use of glucose, oxygen, nitrogen, phosphate and other nutrients when they are available, and sporulate when they are not. The versatility of the catabolic pathways of glucose metabolism is illustrated by reviews of NMR research on yeast in Chapter 10 and yeast spores in Chapter 11. These chapters illustrate the varied responses of glucose pathways as yeast adapt to the extreme range of environments they may naturally experience – conditions which can easily be mimicked under laboratory conditions.

Three states of yeast are reviewed – glucose repressed and derepressed cells in Chapter 10 and the dormant spore in Chapter 11. Yeast's transitions between these states are triggered by changing availability of glucose, oxygen, nitrogen and phosphate. Glucose-repressed cells, growing aerobically in a high glucose environment, represent the simplest case. In such cells, at least in the short term before gene expression changes significantly, oxygenation does not induce any changes in glucose catabolism. In contrast, when derepressed cells are suddenly exposed to high glucose, they go through a transition period of adapting to the nutrient in which gene expression is slowly altered to allow the yeast to better cope with its new environment. During the transition period, of many minutes, yeast are flooded with more glucose than they can oxidize and NMR experiments have measured the pathways of glucose metabolism in the early stages of glycolysis. Under these 'turbo' conditions the cells cycle hexose monophosphates and fructose bis-phosphate back to trehalose, thereby storing carbon and consuming the excess ATP generated by the sudden rich medium. These two upward flows are 'futile cycles', which serve no function in steady-state conditions, but during this transition store excess glucose and dispose of its excess energy – hardly futile activities.

Yeast spores in steady-state dormancy have different requirements for survival in the absence of external nutrients. They must use their stored carbohydrates, trehalose, slowly while at the same time they must supply the energy required for a low level of maintenance. NMR studies of yeast spores in Chapter 11 have shown how the response of the enzyme trehalase to environmental glucose is controlled during dormancy. The control is exercised by ATP feedback and maintains ATP at 0.5 mM. A likely consequence of the ATP levels being 10-fold lower than in yeast is that cellular ATPases that respond to ~5 mM ATP in the vegetative state will be slowed significantly in the spores. This presumably would satisfy survival needs by slowing energy consumption. Although trehalose has been intensely studied as a protector from heat shock, its energetic role as determined by the ^{31}P and ^{13}C experiments has been almost completely neglected in the 20 years since it was described. The importance of energetics in metabolism, a central theme of this book, and, of course, life itself, is emphasized by these simple studies of yeast spores.

13.6. ENERGY AND POWER IN MUSCLE WORK

Three chapters on the muscle and its work are considered together to provide novel interpretations of new and old data and to suggest a hypothesis for future studies. Chapter 7 showed that the millisecond firings of muscle require rapid generation of energy. Chapter 8 proposed that this rapid energy could only be generated by non-oxidative glycolysis. The lactate build-up during exercise is not caused by oxygen deficits (since results show that oxygen is fully available), but lactate serves as a temporal buffer, coupling the rapid power needs of contractions generated anaerobically to the efficient but slow energy available from the oxidation of glucose. Lactate, in this picture, is not a pathological symptom of energy consumption having outstripped supply, but of the millisecond bursts needing rapid power and thereby exceeding the averaged energy requirements.

Three experimental studies are the basis of this picture. First, the lactate shuttle proposed by George Brooks showed that lactate was produced under well-oxygenated conditions and its reducing potential could be further utilized by efflux to other tissues. This then opened the question as to why lactate was generated at all. The answer started with the second study [i.e. the rapid turnover of phosphocreatine (PCr) in exercise shown by millisecond ^{31}P MRS]. The spectra were triggered by an electrical stimulus at a muscle producing a millisecond resolved sequence of ^{31}P spectra and showed the time course of PCr, ATP, P_i and pH. The PCr peak dropped by ~3 mM with a time constant of ~14 ms and recovered back to pre-stimulus levels by ~30 ms. Hence the energy costs of an exercise regime, which is commonly evaluated by the loss of PCr at the end of many contractions, must include the energies continually being used to restore ATP and PCr. The source of rapid non-oxidative energy in the muscle was proposed in Chapter 8 to be glycogen, whose rapid turnover as a result of continuing glycogenolysis and glycogenesis was detected by the third kind of experiment. ^{13}C MRS measurements showed that reduced but constant concentrations of glycogen were maintained during exercise thanks to continual synthesis and degradation of glycogen.

Although both the ^{31}P and ^{13}C MRS experiments described in these chapters require elaboration and experimental follow-ups, they do provide a novel view of a very well studied biochemical pathway involved in energy storage and utilization in muscle work. These experiments open approaches to unanswered questions about muscle physiology such as what is the relation between contraction and energy and what are the mechanisms and causes of muscle fatigue.

NMR studies of glycogen in the heart introduced a revealing perspective on cardiac metabolism and reinforced a warning about the usefulness of mutants. The heart can oxidize different substrates – glucose, ketones, fats or lactate – and the flow of 1-^{13}C glucose into glycogens, measured by NMR, reflected changing fuel consumption. Lactate either infused or increased after epinephrine or hypoxia, diverted glucose into glycogen. Lactate, contributing to the glycolytic flux, increased glucose-6-phosphate, at the branching point of glycogen synthesis and glycolysis. The novel interpretative showed that glucose uptake, controlled by transport flux, was then diverted to glycogen to the extent that its contribution to glycolysis was replaced by lactate. In this simplifying analysis the effects of systemic parameters like epinephrine and hypoxia upon glycogen metabolism were more directly evaluated from the metabolite lactate, changed by these conditions, than by trying to interpret the metabolic consequences of epinephrine.

The hypothetical model of transport control of glucose metabolism received support from the effects upon glycogenesis of three modifications of transporter activity. Two mutants which decrease transporter density were found actually to have increased it because the rodents overcompensated for the loss. A third intervention, which actually did decrease glucose transport, provided additional support for the glucose transport control of glycogen synthesis. Finally these two mutants illustrate problems frequently encountered when using genetic knockout to modify metabolism. The resulting overcompensation showed that the very activity intended to be down-regulated by the mutations was actually increased presenting a paradox rather than an answer.

13.7. EXPLANATION AND VALIDATION OF ^{13}C MRS METHODOLOGY

Two chapters discuss ^{13}C MRS experiments describing methods and examining their validity. Chapter 2 introduces NMR methods for studying the three nuclei most often used for *in vivo* studies. ^{1}H and ^{31}P experiments have been often tested and are described as straightforward, established methods. On the other hand, ^{13}C MRS, which forms the basis of much of this book's research, receives a more detailed discussion. Because the methods of localization and quantitation have to a great extent been developed for the research described herein, there are novel methodological aspects and these are described. Furthermore the surprisingly well-resolved ^{13}C and ^{1}H NMR spectra of the large glycogen molecule raised questions about its degree of visibility and the mechanisms responsible. Because of the central role of glycogen detection in the following chapters, these question are answered thoroughly in Chapter 2.

Isotopomer analysis has become uniquely valuable in ^{13}C NMR spectra, both *in vivo* and in extracts, so that this rapidly developing method is discussed carefully in conjunction with a presentation of its recent results in Chapter 12. Isotopomer analysis is made on extracts from the body. The cases discussed in Chapter 12 were tissue extracts, but similar results are available from the study of metabolites in body fluids. In isotopomer studies NMR measurements are made by standard chemical NMR methods capable of rapidly generating large data sets. Isotopomer analyses in this way approach metabolomics methodology, and with much of the same pathway information as *in vivo* MRS.

13.8. METABOLISM AND/OR METABOLOMICS?

Metabolomics has been conceptualized recently as the next step toward understanding biological function, a unification that can build upon the massive pools of data being generated on genetic material by genomics, and on proteins by proteomics. In its early stages metabolomics is being defined in many different ways. In parallel with the massive data accumulation pioneered by genomics, many advocate that metabolomics should be a collection of a great body of information about metabolism. In some formulations, metabolomics plans to take advantage of the existing information from intermediary metabolism as well as from the chemical pathways studied by molecular biology. However the magnitude of data collection envisaged by the metabolomic approach, while it can be large, must not be confused with completeness, because *in vivo* metabolism is specific to the individual and to the contingent conditions established by individual history and to the present state of subject and environment and their continuing dialectic. Therefore, the means by which it is proposed that metabolomics reaches this particular dream are not possible. A complete measurement of every chemical in a particular cell is only one data point in the large, many-dimensional set of variables and cannot provide an understanding that goes beyond the particular data. Disregarding these limitations, in general metabolomics proposes to collect data on metabolites while interpretations are envisaged to come from the concepts in the biochemistry textbooks that all biologists have read at some time. In this simplistic procedure, there is an implicit faith that the metabolism described in those textbooks is fully developed and represents reserves of static truths. It is suggested that metabolomics will let us develop vast new stores of information about, for example, the concentration of all metabolites in a particular cell or organ, and then, with the new data, acquired with more and more refined instrumental methods, we will turn to the existing, thoroughly understood results and concepts of intermediary metabolism to explain the complex functions of life – a bit like an assembly line of discovery.

Needless to say, we disagree with this approach. We share the dream of relating organismic functions to properties of the molecules of life. Yet, in contrast to proposing that large projects will accumulate information that might eventually serve that goal while assuming that the theoretical understanding is already in hand, we have shown metabolic studies that have already made such connections. Metabolic concepts are being revised by the new methods described in this book and offer hopes for even smoother integration of molecular and organism function. Metabolism does have a vast store of understanding, but its meaningfulness is not static. Rather, the dynamic aspects of metabolic research show how the newer results and concepts obtained by *in vivo* NMR, and guided by metabolic control analysis, presently allow metabolic research to connect molecules with the higher activities of the human body.

Index

Note: Figures and Tables are indicated by *italic page numbers*, notes by suffix 'n'

Metabolomics by In Vivo NMR. Edited by R. G. Shulman and D. L. Rothman

ISBN: 0-470-84719-0

With kind thanks to Paul Nash for creation of this index.